A Special Issue of
Social Neuroscience

Neural Correlates of Deception

Guest Editors

Giorgio Ganis
Harvard Medical School/Martinos Center for Biomedical Imaging, USA

and

Julian Keenan
Montclair State University, USA

Psychology Press
Taylor & Francis Group
HOVE AND NEW YORK

First published 2010 by Psychology Press
27 Church Road, Hove, East Sussex, BN3 2FA

www.psypress.com

Simultaneously published in the USA and Canada
by Psychology Press
270 Madison Avenue, New York, NY 10016

Psychology Press is part of the Taylor & Francis Group, an informa business

British Library Cataloguing in Publication Data
A catalogue record for this book is available from the British Library

ISBN 13: 978-1-84872-712-0
ISSN 1747-0909

Cover design by Design Deluxe
Typeset in India by Datapage International
Printed in the UK by Hobbs the Printers Ltd, Totton, Hampshire
Bound in the UK by TJ International Ltd, Padstow, Cornwall
The publication has been produced with paper manufactured to strict environmental standards and with pulp derived from sustainable forests.

Contents*

The cognitive neuroscience of deception
Giorgio Ganis and Julian Paul Keenan 465

Limitations to the detection of deception: True and false recollections are poorly distinguished using an event-related potential procedure
John J. B. Allen and Ralf Mertens 473

Thirty-site P300 scalp distribution, amplitude variance across sites, and amplitude in detection of deceptive concealment of multiple guilty items
Ming Ann Lui, J. Peter Rosenfeld, and Andrew H. Ryan Jr. 491

Can simultaneously acquired electrodermal activity improve accuracy of fMRI detection of deception?
F. Andrew Kozel, Kevin A. Johnson, Steven J. Laken, Emily L. Grenesko, Joshua A. Smith, John Walker, and Mark S. George 510

Towards clinical trials of lie detection with fMRI
J. G. Hakun, K. Ruparel, D. Seelig, E. Busch, J. W. Loughead, R. C. Gur, and D. D. Langleben 518

Detection of deception using fMRI: Better than chance, but well below perfection
George T. Monteleone, K. Luan Phan, Howard C. Nusbaum, Daniel Fitzgerald, John-Stockton Irick, Stephen E. Fienberg, and John T. Cacioppo 528

Neural processes underlying self- and other-related lies: An individual difference approach using fMRI
Giorgio Ganis, Robert R. Morris, and Stephen M. Kosslyn 540

Types of deception revealed by individual differences in cognitive abilities
Charity J. Morgan, Julia B. Lesage, and Stephen M. Kosslyn 555

The effect of deception on motor cortex excitability
Karen J. Kelly, Elizabeth Murray, Veronica Barrios, Jamie Gorman, Giorgio Ganis, and Julian Paul Keenan 571

Subject index 575

---

* This book is also a special issue of the journal *Social Neuroscience*, and comprises pp. 465–574 of Volume 4 (2009). The page numbers are taken from the journal and so begin with p. 465.

SOCIAL NEUROSCIENCE, 2009, 4 (6), 465–472

Psychology Press
Taylor & Francis Group

The cognitive neuroscience of deception

Giorgio Ganis
Harvard Medical School, Boston, MA, USA, Massachusetts General Hospital, Charlestown, MA, USA, and Harvard University, Cambridge, MA, USA

Julian Paul Keenan
Montclair State University, Montclair, NJ, USA

INTRODUCTION

William Hazlitt (1778-1830), a British writer, once asserted that, "life is the art of being deceived." Human social relations are so steeped in deception that it is impossible to imagine life without it. From great drama sweeping one away momentarily to interleaved complex romantic relationships, the joining of humans is cemented in place by deception. The basis of who we are (or aren't) in relation to others is often predicated on falsification such that all moral, legal, and ethical systems must take account of this core feature of human nature. Some researchers have gone so far as to postulate that human brains are innately primed to deceive, since deception is recorded in all societies, extending back to the earliest written record, and it occurs early in life in a predictable manner. Given the appropriate abstract reasoning skills, along with basic social abilities, human brains quickly discover that rewards outweigh the costs associated with deception. It is therefore an important task of social neuroscience to reveal the inner workings of deception.

Deception can be defined in more than one way. According to some definitions, the intentional manipulation of the truth is not an essential feature of deception. Such definitions would include, for example, the behavior of animals that mimic or blend into their background, in order to escape detection by predators. Broader definitions would include cases such as self-deception, where the presence of another individual is not even required.

For our purposes, we will define deception as a social behavior in which an individual attempts to persuade another to accept as true what the deceiver believes to be untrue. Although in our culture deception usually has negative connotations, it is a fundamental and pervasive social behavior with an important role in our daily life. For example, research has shown that most of us lie to others many times each day, usually for the purpose of managing social relationships. Conversely, most of us frequently attempt to determine whether others are lying to us (e.g., DePaulo, Kashy, Kirkendol, Wyer, & Epstein, 1996).

The simplest and most immediate way to try to find out if someone is lying consists in looking at their behavior. It is commonly believed, for instance, that people tend to look more nervous when they lie than when they tell the truth. This is why, initially, the study of deception relied on the observation of behavioral cues thought to accompany the generation of deceptive responses. Such cues were postulated to be caused by increased physiological arousal due, for example, to fear of getting caught or increased cognitive load when generating deceptive responses. It has been established that there are correlations between deception and behavioral cues; for instance, liars tend to fidget more than non-liars (DePaulo et al., 2003), but these correlations are not strong and cannot be used as diagnostic features of deception.

Correspondence should be addressed to: Giorgio Ganis, Massachusetts General Hospital, 149 Thirteenth Street, Charlestown, MA 02129, USA. E-mail: ganis@nmr.mgh.harvard.edu

www.psypress.com/socialneuroscience DOI:10.1080/17470910802507660

Attempting to find better correlates of deception by measuring arousal more directly, researchers began monitoring peripheral psychophysiological variables such as heart rate, breathing patterns, and electrodermal activity, the main measures used by the polygraph. Although peripheral psychophysiological measures are still widely used to study and to detect deception, the link between deception and arousal has been questioned because individuals can be aroused for reasons that have little to do with deception, and because not all individuals are aroused when they produce deceptive responses.

Because of these problems, researchers have begun to study directly the brain, the organ that produces the lies, with the goal of finding more reliable correlates of deception than has been possible with measures of arousal. During the last two decades, cognitive neuroscience has provided important tools required for this type of investigations. Since the late 1980s, researchers have used event-related potentials (ERPs) to study deception. For example, some investigators have exploited the fact that recognition of infrequent and familiar events (e.g., crime scene details) modulate brain potentials such as the P300 (e.g., Rosenfeld et al., 1988) or that response conflict (such as that created by having to inhibit an honest response while producing a deceptive one) modulates the amplitude of medial frontal negativities (e.g., Johnson, Barnhardt, & Zhu, 2004). In the past few years, researchers have begun using functional magnetic resonance imaging (fMRI) with variants of classic deception paradigms to attempt to determine more directly the neural correlates of deception (for brief reviews see Greely & Illes, 2007; Spence et al., 2004). At the same time, the relationship between these neural measures and more traditional measures of deception has begun to be explored.

This special issue provides a representative sample of new laboratory research on the cognitive and neural processes associated with the generation of deceptive responses using a variety of paradigms, techniques, and measures. Before summarizing the contents of this issue, we will briefly outline a subset of current topics in deception research.

TOPICS IN DECEPTION RESEARCH

Producing a deceptive response is a complex process that usually involves: (a) Deciding to lie, after evaluating the full social and pragmatic context of the situation; (b) retrieving information associated with honest and potentially deceptive responses from memory; (c) maintaining and manipulating relevant retrieved information in working memory (at least until a response is made); (d) encoding into memory of the response itself in order to maintain consistency over time; (e) inhibiting alternative deceptive responses and the corresponding honest response, coincidentally retrieved. Recruiting these cognitive processes is logically necessary to produce a deceptive response, regardless of whether an individual is anxious or feels guilty about lying, although the strength of engagement may be modulated by these factors. Most studies have focused on interference monitoring and resolution processes required to inhibit honest responses, conceptualized as "prepotent" responses (Spence et al., 2004), when producing deceptive ones. Understanding the interplay between these processes and how they relate to other cognitive processes is crucial for any theory of deception.

Group versus single individual data

Historically, studies on deception have revolved around two themes and have focused on two partially interrelated goals. The first goal involves understanding the cognitive and neural correlates of deception and their relationship with other processes, in order to develop general theories of deception. To achieve this goal, one usually only needs to demonstrate significant differences between experimental conditions in group data. The second goal, more applied, is about whether it is possible to achieve reliable detection of deceptive responses. To achieve this goal, it is necessary to be able to discriminate between deceptive and honest responses in single individuals and, ideally, in single trials. This is an especially challenging task, because the signal-to-noise ratio of the single individual data during high-level cognitive tasks such as deception is usually rather low. In order to increase the signal-to-noise ratio, researchers have been trying to combine measures, either from different modalities (e.g., electrodermal and hemodynamic measures) or from different channels within the same modality (e.g., signals from different brain regions).

Different types of lies and individual differences

Two sources of variability in deception research are often acknowledged, but they have not been explored systematically, especially with neuroimaging. These sources of variability may account for some of the puzzling inconsistencies among different studies. For example, different parts of medial frontal cortex have been found by contrasting deceptive and honest responses in different neuroimaging studies. The first source of variability derives from the fact that there are different types of lies, which are likely to be associated with different neural processes. For instance, lies that are prepared in advance and rehearsed engage brain networks that are partially nonoverlapping with those engaged by spontaneous lies that are generated on the fly (Ganis et al., 2003). The second source of variability comes from individual differences. For example, it has been reported that individuals who score high on the alienation subscale of the Psychopathic Personality Inventory tend to show stronger activation in medial frontal regions during deception (Nunez, Casey, Egner, Hare, & Hirsch, 2005). Understanding how these sources of variability contribute to the cognitive and neural correlates of deception is a critical area of research.

Laboratory versus field studies of deception

Mostly due to practical reasons, deception paradigms used in laboratory settings, especially if they involve neuroimaging measures, are impoverished compared to real-life situations (e.g., Greely & Illes, 2007; Sip, Roepstorff, McGregor, & Frith, 2008), raising classic issues about the relative merits of ecological versus laboratory approaches (e.g., Banaji & Crowder, 1989; Neisser, 1978). There is no a priori reason to believe that laboratory studies of deception will be less informative than they have been in other fields (e.g., vision research, where, for example, participants fixate stimuli presented on computer screens instead of freely viewing visual stimuli in the real world). However, two issues are worth discussing to place the contributions to this special issue, all laboratory studies, in perspective.

First, in laboratory studies the *motivation to lie* is generally low. On the one hand, typical laboratory paradigms are based on lying about events such as what playing card participants hold. On the other hand, even with various mock-crime and competitive game scenarios, participants are generally aware that the situation is artificial, and that there are no important consequences (positive or negative) resulting from their actions.

Second, for experimental design reasons, participants are *instructed* to lie (implicitly or explicitly), which, of course, is not the case in real-life deception situations. The impact of these two factors on deception processes remains to be established by future studies, and it may be different for the two research goals mentioned earlier. On the one hand, it is possible that realistic deception situations where people are bound to lose social capital (e.g., being labeled as untrustworthy) engage brain networks that are recruited only weakly, or not at all, in laboratory situations that involve lying to the experimenter in a cooperative situation. Even if this were true, however, numerous processes studied in the laboratory are likely to be required also during deception in real-life situations. For example, when lying, one needs to inhibit honest responses both in the laboratory and in real-life situations. On the other hand, if it is still possible to detect deception in single individuals under rather impoverished laboratory conditions, it is possible that adding additional dimensions typical of real-life situations could provide further dimensions that can be used to better discriminate between deceptive and honest responses.

The contributions to this special issue address these topics to various degrees.

CONTENTS OF THE SPECIAL ISSUE

It may be convenient to categorize the studies in this special issue according to two methodological dimensions, the paradigm and dependent variables employed. Regarding the paradigms, a subset of the studies used variants of the Guilty Knowledge Test (GKT), also referred to as the Concealed Knowledge Test (CIT). The basic paradigm relies on the idea that the perpetrators of a crime would exhibit signs of recognition when they see an item from the crime scene (usually referred to as "probe" item), even if they lie and deny recognizing it, relative to other items that could have plausibly been present at the crime scene but were not (usually referred to as "irrelevant" items). In contrast, innocent individuals would respond equally to all items, including the one from the crime scene, since they are equally unfamiliar with all of them. Besides providing a tool to investigate the neural correlates of some types of deception, this paradigm is thought by prominent researchers in the field to have the potential to become admissible in criminal courts (e.g., Ben-Shakhar, Bar-Hillel, & Kremnitzer, 2002). Note that the paradigms used here are different from the traditional CIT because people are instructed to lie on the probe items, whereas in the traditional CIT items are processed incidentally and participants are simply told to indicate lack of recognition for all items. A second subset of studies employed variants of the differentiation of deception paradigm (Furedy, Davis, & Gurevich, 1988), in which participants are instructed to lie on half the items and tell the truth on the other half, in some cases with the same items used in both conditions for comparison.

With regard to the dependent variables employed, one study used behavioral measures alone, two studies used electrophysiological measures, three studies used blood-oxygen-level dependent (BOLD) measures obtained with fMRI (some in combination with behavioral measures), one study used electrodermal and BOLD measures, and one study used a measure of motor cortical excitability obtained with transcranial magnetic stimulation (TMS). In these studies, participants were instructed to deceive another person, and they would usually gain money for lying successfully.

Allen and Mertens used ERPs to address a version of an important question in deception research. Can we distinguish between deceptive responses and honest responses that convey false information because individuals falsely remember the details of an event? Technically in this case, people are not lying, based on the above definition of deception, because they are not intentionally trying to mislead another person. Specifically, as mentioned earlier, GKT paradigms rely on the logic that the perpetrator of a crime will recognize details of the crime scene, and distinguish them from other details that were not present in the scene, whereas innocent individuals will not. However, one assumption of this logic is that the information in memory reflects closely the details of the crime scene. Therefore, known memory distortions may be especially problematic for deception detection methods based on recognition. Two studies addressed this issue by determining whether ERPs can distinguish between true and false recognition using the Deese-Roediger-McDermott (DRM) paradigm to elicit false memories. In this paradigm, studying a list of words (e.g., bed, rest, blanket) that are strongly associated with a critical item that is never presented (e.g., sleep), results in high rates of false recognition of the critical item in subsequent recognition tests. Thus, one can compare ERPs to falsely and correctly recognized items. The results indicated that ERPs, in particular P300 amplitude and latency, could not distinguish between true and false memories during the recognition test. This suggests that in some cases, both true and false memories elicit similar ERPs that index recognition processes. Although it remains to be seen whether these results hold with more realistic scenarios, the findings show the type of difficulties that need to be overcome before one can use these methods in real-life situations.

A different reason why the neural processes associated with deceptive responses may look like those associated with honest responses is due to the use of countermeasures. Countermeasures are defined as behaviors used by participants to try to defy deception detection. The extent to which countermeasures can alter the signatures of deceptive responses, making them look like honest responses, is a key issue for potential applications of deception research. It has been shown that cognitive countermeasures can dramatically reduce the classification accuracy based on ERP measures during GKT tasks (e.g., Rosenfeld, Soskins, Bosh, & Ryan, 2004). For example, explicitly associating covert actions (e.g., imagined movements of a specific finger) with irrelevant items during a GKT task can increase the salience of these stimuli and make the amplitude of the P300 they elicit virtually indistinguishable from that elicited by the probe items. However, most ERP studies on the topic have investigated only the amplitude of the P300 to carry out the discrimination. Lui, Rosenfeld and Ryan investigated whether there are features in the spatial distribution of the P300 over many recording sites (variance of P300 amplitude over recording sites) that can be used to discriminate deceptive from honest responses in a GKT task. The basic idea is that differences in the spatial distribution of the P300 may be more robust than simple P300 peak amplitude to countermeasures. The results showed that P300 distribution variance methods contain information that can be used to discriminate deceptive and honest responses, independently of peak amplitude. In principle, this kind of information could be employed to distinguish deceptive and honest responses in the presence of countermeasures.

These types of analyses are an example of the use of multidimensional measures to attempt to capture more complex aspects of deception processes and to achieve more robust discrimination between deceptive and honest responses.

The use of fMRI, with the intrinsic multidimensionality of neuroimaging datasets, as well as the combination of data from various techniques, goes further in this direction. Kozel and collaborators followed up on their fMRI research which showed that they could discriminate deceptive from honest participants in more than 90% of the cases by using brain activation from 3 regions of interest (right orbitofrontal, right middle frontal, and right anterior cingulate cortex) defined in a separate group of participants. They used the electrodermal measures that were collected simultaneously with fMRI in their original study, in order to compare peripheral measures of arousal with brain-related measures of deception. The results showed that the electrodermal responses correlated with activation in orbitofrontal and anterior cingulate cortex, consistent with prior work suggesting a role for these regions in interfacing cognitive and autonomic systems. Furthermore, these regions were different from those found in the contrasts between deceptive and honest responses, indicating that BOLD and electrodermal measures tap into different aspects of deception processes, at least in paradigms with low stakes and fast rates of stimulus presentation. Importantly, including the electrodermal activity data as a further dimension in the discrimination analyses did not appear to increase the accuracy rates, relative to BOLD measures alone. More studies such as this one are required to compare multiple dependent measures of deception and to understand their dependence on the details of the deception paradigms. For instance, electrodermal measures may work better with more realistic situations and higher stakes, as suggested by polygraphic studies.

In an attempt to create a more realistic deception situation, Hakun and colleagues conducted a single case study with fMRI, using a variant of the GKT, referred to as the Resume Query Test (RQT). In this paradigm, the participant (a *Washington Post* reporter, who volunteered for the study) was instructed to pretend to be applying for a job as a science writer and to employ "whatever means necessary to be hired." These instructions were meant to encourage the participant to falsify some items on his resumé, without instructing him explicitly to lie. During the fMRI session, the participant was asked questions and shown pictures about three resumé items that could be easily verified (known truth) and about three items that could not be easily verified (unknown, which the participant confirmed were lies, after the study). Comparing activation to each unknown item with the combined known truth items showed differences in a number of regions that had been revealed in previous studies, including the inferior frontal and medial frontal gyrus. However, the results varied by item. For example, deceptive responses to the question "Did you serve in the military?" showed activation in the medial temporal lobe, which was not seen for the other deceptive responses. Although replication of this single case study is needed, these results highlight the substantial item-to-item variability present in complex tasks such as deception and the need to devise analytic techniques to capture such variability, for both theoretical and applied reasons.

A related issue, with theoretical and practical implications, is the extent to which the networks of areas engaged by deception vary from person to person. Monteleone and collaborators followed up on their neuroimaging work and addressed one version of the individual variability question by determining how well they could discriminate deceptive and honest responses in single individuals by using regions of interest defined by group results. In their study, they used a jackknife method in which the results of the group analyses including $N-1$ participants were employed to detect deception in the left-out, *N*th, individual. The findings indicate that no single region could discriminate deceptive and honest responses accurately in all individuals. The region with the best discrimination rates across individuals was a portion of the medial prefrontal cortex, possibly reflecting response inhibition processes, which could correctly classify 71% of individuals as lying, without false alarms. These results suggest that analyses taking into account individual differences may be required to better characterize deception processes and, potentially, to improve classification accuracy.

Ganis, Morris, and Kosslyn extended their previous fMRI work on the neural correlates of different types of lies by investigating individual differences as a function of type of lie. Participants produced two types of deceptive responses, lies about oneself (self-related) and lies about other individuals (other-related), in addition to the corresponding honest responses. Trial types were randomly intermixed, and participants were instructed whether to lie or not by means of a brief cue, presented at the beginning of each trial. The effect of individual variability was assessed by introducing the notion of rate-limiting process. Rate-limiting processes for a task act as performance bottlenecks for this task. Using a simple analogy to illustrate, finger strength is a rate-limiting process for the task of opening jars, since people with stronger fingers tend to perform this task faster and more accurately than people with weaker fingers. However, finger strength is not a rate-limiting process for other tasks, such as typing on a computer. Brain regions that implement rate-limiting processes for the deception tasks used in the study were uncovered by finding brain regions in which activation predicted performance. One prediction was that a cingulate-insular-prefrontal network, thought to be rate limiting for interference tasks, would also be involved in deception because of the interference produced by the prepotent honest response when producing a deceptive response. The main finding confirmed this prediction and also showed an intriguing pattern of dissociations between the two types of lies. First, partially non-overlapping sets of rate-limiting regions were found for self- and other-related lies. Second, similar cingulate-insular-prefrontal networks were found for both types of lies; however, the correlation between brain activation and response times was positive for self-related lies but negative for other-related lies. This pattern of dissociation was interpreted in terms of different characteristics of self- and other-related lies.

A similar logic to exploit individual variability, but using behavioral measures, was employed by Morgan, Lasage and Kosslyn. In this study, a battery of cognitive tasks was administered to all participants performing the deception tasks used in previous work. The general logic of this study was to determine whether performance on different subsets of cognitive tasks predicted performance on different types of lies. Specifically, if a subset of the cognitive tasks tap into processes that are rate-limiting for some of the deception tasks, then performance on these cognitive tasks should correlate with performance on the deception tasks, across individuals. Prior work has shown that the production of spontaneous lies engages the anterior cingulate and adjacent medial prefrontal cortex, compared to honest responses, but the production of rehearsed lies does not. This suggests that spontaneous lies recruit interference monitoring and resolution processes more strongly than rehearsed lies. Accordingly, one interesting prediction of the study was that performance on tasks thought to recruit interference monitoring and resolution processes (for example, variants of the Stroop task) should predict performance on the production of spontaneous but not rehearsed lies. The results confirmed this prediction, indicating that this methodology may be promising in breaking down deception processes into subcategories.

Finally, Kelly and collaborators investigated interference resolution and response inhibition processes during deception by recording motor evoked potentials induced by TMS to measure motor cortical excitability. In a clever deception paradigm, sport fans were shown pictures of their preferred team as well as of the opposing team and they were asked to respond deceptively or honestly about which team they preferred. The main question addressed in this study was whether there would be consistent differences in cortical excitability (in left and right motor cortex) between deceptive and honest responses. The results indicated a dissociation in which the excitability of the left motor cortex is increased during deception, whereas that of right motor cortex is decreased. These results complement neuroimaging data showing increased left motor cortex activation during deception, and show the promise of this kind of technique in the study of deception.

FUTURE DIRECTIONS IN DECEPTION RESEARCH

On both theoretical and practical levels, the study of the neural correlates of deception examined here represents only a small sampling of what the field of the future will look like. The contributions to this special issue raise a number of questions and suggest some future directions for deception research.

First, a key question is whether deception processes are special in any way, or whether they rely on a set of general-purpose processes. As discussed, a number of studies have postulated a connection between interference monitoring and resolution processes and the production of at least some types of deceptive responses. Yet there has not been conclusive evidence on whether deception relies on a special kind of interference monitoring and resolution processes, perhaps specialized for managing the type of social negotiations usually required by deception. More research will be needed to answer this theoretically important question.

Second, given the broad range of paradigms and techniques used to study deception, another important question is how these methods relate to each other and how they can be integrated. Do different measures tap into different aspects of deception processes? Are they affected by different factors? For example, stakes may affect electrodermal measures more than they affect hemodynamic or electrophysiological measures. Similarly, factors that affect the amplitude of the P300 may not affect the amplitude of hemodynamic responses observed with the same paradigms. Progress in this direction is likely to have important theoretical and practical consequences.

Third, it is clear that deception is not a unitary phenomenon and that much of the future work, both theoretical and empirical, will involve determining more systematically the dependence of deception processes on key variables, including the type of lie and the social and pragmatic context in which lying takes place. On the one hand, this will enable researchers to fine-tune theories of deception. On the other hand, this will be important for applied deception research because one could develop more accurate deception detection methods that are optimized for specific domains. Creating a collection of such methods may be more productive than attempting to devise a general-purpose deception detection procedure.

Fourth, although it may be nearly impossible to study real-life situations with neuroimaging, at least currently, efforts towards designing paradigms that approach such situations are important. It may turn out that the main results of laboratory studies are not very different from those of ecological studies, but a complete theory of deception should be able to handle factors that may come into play only in real-life situations, for example the fact that in real-life situations people typically decide when to lie.

The rapidly growing field of social neuroscience will clearly move in unknown directions in the future. However, given the importance of deception in our social lives, it is likely that deception will receive greater attention in the upcoming years. Establishing the brain correlates of deceptive behaviour will be a difficult task, but already we have seen great strides. We are excited to see what comes next.

REFERENCES

Banaji, M., & Crowder, R. (1989). The bankruptcy of everyday memory. *American Psychologist*, *44*, 1185–1193

Ben-Shakhar, G., Bar-Hillel, M., & Kremnitzer, M. (2002). Trial by polygraph: Reconsidering the use of the guilty knowledge technique in court. *Law and Human Behavior*, *26*(5), 527–541.

DePaulo, B. M., Kashy, D. A., Kirkendol, S. E., Wyer, M. M., & Epstein, J. A. (1996). Lying in everyday life. *Journal of Personality and Social Psychology*, *70*(5), 979–995.

DePaulo, B. M., Lindsay, J. J., Malone, B. E., Muhlenbruck, L., Charlton, K., & Cooper, H. (2003). Cues to deception. *Psychological Bulletin*, *129*(1), 74–118.

Furedy, J. J., Davis, C., & Gurevich, M. (1988). Differentiation of deception as a psychological process: A psychophysiological approach. *Psychophysiology*, *25*(6), 683–688.

Ganis, G., Kosslyn, S. M., Stose, S., Thompson, W. L., & Yurgelun-Todd, D. A. (2003). Neural correlates of different types of deception: An fMRI investigation. *Cerebral Cortex*, *13*(8), 830–836.

Greely, H. T., & Illes, J. (2007). Neuroscience-based lie detection: The urgent need for regulation. *American Journal of Law and Medicine*, *33*(2-3), 377–431.

Johnson, R., Jr., Barnhardt, J.T. & Zhu, J. (2004). The contribution of executive processes to deceptive responding. *Neuropsychologia*, *42*(7), 878–901.

Neisser, U. (1978). Memory: What are the important questions. In M. Gruneberg, P. Morris, & R. Sykes (Eds.), *Practical aspects of memory*. New York: Academic Press.

Nunez, J. M., Casey, B. J., Egner, T., Hare, T., & Hirsch, J. (2005). Intentional false responding shares neural substrates with response conflict and cognitive control. *Neuroimage*, *25*(1), 267–277.

Rosenfeld, J. P., Cantwell, B., Nasman, V. T., Wojdac, V., Ivanov, S., & Mazzeri, L. (1988). A modified, event-related potential-based guilty knowledge test. *International Journal of Neuroscience*, *42*(1-2), 157–161.

Rosenfeld, J. P., Soskins, M., Bosh, G., & Ryan, A. (2004). Simple, effective countermeasures to P300-based tests of detection of concealed information. *Psychophysiology, 41*(2), 205–219.

Sip, K. E., Roepstorff, A., McGregor, W., & Frith, C. D. (2008). Detecting deception: The scope and limits. *Trends in Cognitive Sciences, 12*(2), 48–53.

Spence, S. A., Hunter, M. D., Farrow, T. F., Green, R. D., Leung, D. H., Hughes, C. J., et al. (2004). A cognitive neurobiological account of deception: Evidence from functional neuroimaging. *Philosophical Transactions of the Royal Society of London. Series B, Biological Sciences, 359*(1451), 1755–1762.

SOCIAL NEUROSCIENCE, 2009, 4 (6), 473–490

Psychology Press
Taylor & Francis Group

Limitations to the detection of deception: True and false recollections are poorly distinguished using an event-related potential procedure

John J. B. Allen and Ralf Mertens
University of Arizona, Tucson, Arizona, USA

The utility of using indices of neural function to identify deception relies on finding highly reliable and valid approaches that adequately identify the guilty and exonerate the innocent. A class of approaches, based on the guilty knowledge technique (GKT), assume that guilty individuals will recognize specific crime-relevant details, whereas innocent individuals will not. Memory distortions, however, may limit the accuracy of such procedures. To investigate these limits, two studies were conducted to examine whether brain electrical activity could differentiate true from false recollections elicited by the Deese-Roediger-McDermott (DRM) paradigm. The design of each study maximized the opportunity of finding electrocortical differences between true and false recognition. Each study found very high rates of false recognition, with little evidence that brain electrical activity could differentiate true from false memories. Results suggested that under certain conditions both true and false recollections can produce a pattern of brain activity indicative of recognition.

INTRODUCTION

Most of us are notoriously poor at detecting deception (e.g. Ekman and O'Sullivan (1991), although see O'Sullivan & Ekman (2004) for accounts of a few detection "wizards"), which has inspired humans to devise numerous techniques that may assist in the detection of deception. No technique can detect lying or deceiving *per se*; instead these techniques measure other processes that may be correlated with deception; e.g., assessing the anxiety or arousal that accompanies the fear of being caught lying, or assessing memory of events for which only a guilty individual might have knowledge. Technologies range from the use of primitive props (e.g., assessing dry mouth using masticated rice in ancient China, Adler (2007)) to implementing sophisticated modern neuroimaging technologies to assess brain activity during lying (e.g., Ganis, Kosslyn, Stose, Thompson & Yurgelun-Todd, 2003; Kozel, Johnson, Mu, Grenesko, Laken, George, 2005; Langleben et al., 2005). Because such techniques do not measure lying directly, but instead processes correlated with lying, each will have vulnerabilities that could render the technique undesirably inaccurate in certain situations, which may lead to false accusations of guilt among innocent individuals, or false exoneration of guilty individuals.

Table 1 provides a list of three broad approaches to the detection of deception, with illustrative examples of each approach. One of the most widely used, and also most widely

Correspondence should be addressed to: John J. B. Allen, Department of Psychology, University of Arizona, PO Box 210068, Tucson, AZ 85721-0068, USA. E-mail: jallen@u.arizona.edu

The authors thank Veronica Rodriguez, Jeffrey Chang and Lori Anne Tracy for their invaluable help in recruiting and running participants for these studies. Portions of these data were presented at the annual meetings of the Society of Psychophysiological Research in Granada, Spain, October 1999 and San Diego, CA, October 2000.

www.psypress.com/socialneuroscience DOI:10.1080/17470910802109939

TABLE 1
Approaches to detecting deception

Emotion/Arousal	*Memory/Recognition*	*Other cognitive or neural correlates*
The polygraph test	Guilty knowledge test using:	Response conflict
Facial expression	Autonomic (skin conductance response)	Attention and memory load
Voice stress	Central (ERP, fMRI)	Lingusitic analysis
Facial blood flow	Memory malingering tests	Most fMRI approaches
Facial thermography		
Demeanor		

criticized, approaches is the conventional field polygraph examination based on the control-question technique (CQT), the shortcomings of which have been widely noted in the scientific literature (Iacono, 2000; Iacono & Lykken, 1997; Iacono & Patrick, 1997; Lykken, 1987; National Research Council, 2003; Office of Technology Assessment, 1983). The poor validity of this and other arousal-based techniques (Iacono & Patrick, 2006) has inspired alternative methods for the detection of deception, dating back to the introduction of the guilty knowledge technique (GKT; Lykken, 1959), which assesses recognition for key facts rather than arousal that accompanies lying. The GKT can utilize autonomic (e.g., skin conductance) or central nervous system measures (Bashore & Rapp, 1993; Rosenfeld, 1995), including measures of cortical activity such as event-related brain potentials (ERPs). Unlike conventional polygraph approaches using the CQT that assess emotional arousal associated with lying, ERP-based alternatives most often utilize the GKT approach to assess memory for salient aspects of a situation that would only be known to perpetrators and few others such as investigators or witnesses. By contrast, the most recent neuroimaging approaches have utilized functional magnetic resonance imaging (fMRI) to identify brain areas that differ in blood flow during lying compared to truth telling, identifying regions that are important in the complex phenotype of lying, but not identifying a specific brain region uniquely related to lying *per se*.

The present studies focused on using ERPs to assess memory in a GKT format, testing whether the fallibility of memory may render this promising approach invalid for the detection of deception. The GKT assumes that crime-relevant information known only to a guilty individual will be recognized by that individual but not others, and that it can be reliably differentiated from superficially similar but unfamiliar material. Unfortunately, memory is fallible, and individuals may come to believe that certain events occurred that in fact did not, including, for example: (1) accused suspects confusing a fabricated series of events with the original set of events, for example through repeated questioning, leading questions, or rehearsal of an alibi; (2) accusers falsely recalling events that they then accuse a defendant of having committed. In either case, it would be very useful to be able to have a reliable method of differentiating these false from true recollections. Thus the present studies investigated cases where individuals come to falsely believe information though a laboratory manipulation, to determine whether ERP procedures previously shown to reliably and validly identify recognized items would also detect as recognized items that subjects falsely remembered having heard in the context of the experiment.

Memory as constructive and reconstructive

Memories are indeed malleable, a product of active construction and reconstruction, influenced by our tendency to make sense out of the world we perceive (Bartlett, 1932; Loftus, 1981; Neisser, 1966), and by our use of scripts (Schank, 1982). Errors of memory involve both omissions and fabrications, the latter of which are the focus of this report. Specifically, two studies were conducted to determine whether ERPs could differentiate veridical from false memories, in cases where subjects report with conviction recognizing items they had not in fact previously heard.

False recollections occur under a variety of circumstances. People claim to remember fictitious events that arose during leading and suggestive interviewing (Loftus, 1986, 1993; Loftus & Palmer, 1974; Sattler, 1998; Schacter, 1996); they describe events that could never possibly have occurred (Loftus & Pickrell, 1995; Spanos, 1994; Spanos, Burgess, & Burgess, 1994; Spanos, Burgess,

Burgess, Samuels, & Blois, 1999); they modify stories to fit personal views (Bartlett, 1932, pp. 66–76); or they remember details that were never presented to them (Garry, Manning, Loftus, & Sherman, 1996; Norman & Schacter, 1997; Roediger, Jacoby, & McDermott, 1996; Underwood, 1965). Research investigating means to differentiate true from false memories might prove useful in these contexts, and may also illuminate mechanisms of memory that lead to false recollections.

A particularly convenient paradigm for inducing false recollections in the laboratory was developed by Deese (1959), and enhanced by Roediger and McDermott (1995). Investigations utilizing the Deese/Roediger and McDermott paradigm (DRM) found that a single presentation of a list of strong associates to a critical (non-presented) item was sufficient to elicit predictable intrusions of the critical word in subsequent free recall, and at impressively high rates. For example, the serial presentation of bed, rest, awake, tired, dream, wake, snooze, blanket, doze, slumber, snore, nap, peace, yawn, drowsy (learned items) will reliably elicit the critical associate *sleep* in a subsequent recall or recognition task. This strong false recognition effect is remarkable since it occurs under testing conditions that tend to promote accurate recall, such as the use of simple stimuli, a short retention interval, and instructions tapping metacognition (Roediger, 1996).

Even though the possibility of study/test modality (Gallo, McDermott, Percer & Roediger, 2001; McKone & Murphy, 2000) and stimulus presentation effects has been noted (Brainerd, Wright, Reyna & Mojardin, 2001; Fabiani, Stadler, & Wessels, 2000; Hicks & Marsh, 2001), false recognition tends to be robust over time (Brainerd et al., 2001; McDermott, 1996; Payne, Elie, Blackwell & Neuschatz, 1996), does not disappear even after multitrial/test exposure (Budson, Daffner, Desikan & Schacter, 2000; McDermott, 1996, Zaragoza & Mitchel, 1996), and occurs even if participants are explicitly informed about the false recognition phenomenon (McDermott, 1996; McDermott & Roediger, 1998). Under certain conditions, recognition of lure words can even exceed that of learned words (Brainerd & Reyna, 1998; Brainerd et al., 2001).

Event-related potentials, memory, and deception

In the service of assessing memory, ERPs may be a promising approach. Multiple ERP components have been found sensitive to the recollection of past experiences. The P3 component, in particular, has been associated with memory and context updating (Donchin & Coles, 1988) and can be elicited in the absence of an overt response by a participant as long as the stimulus presented is attended to (Mertens & Polich, 1997). ERPs have been shown to be a valid tool in differentiating learned from unlearned words, with recognized words producing larger positive going waveform amplitudes beginning about 300–400 ms after stimulus onset (and typically extending several hundred milliseconds) as compared to unlearned and non-recognized learned items, respectively (e.g., Allen, Iacono, & Danielson, 1992; Paller, Kutas, & McIsaac, 1995; Schnyer, Allen, & Forster, 1997; Senkfor & Van Petten, 1998; van Hooff, Brunia, & Allen, 1996; for review see Friedman & Johnson, 2000). The P3 has been successfully employed in a limited number of ERP-based deception detection studies (Allen et al., 1992; Farwell & Donchin, 1991; Farwell & Smith, 2001; Rosenfeld, Angell, Johnson, & Qian, 1991; for review see Allen, 2002). These studies correctly identified individuals in 89% (Rosenfeld et al., 1991), approximately 90% (Farwell & Donchin, 1991) and 95% (Allen et al., 1992) of cases, utilizing different statistical approaches (bootstrapping of peak-to-peak amplitude, bootstrapping of cross-correlations, and Bayesian analysis, respectively). Additionally, a series of studies by Rosenfeld and colleagues have investigated ERPs as a means of detecting malingered or simulated amnesia, with the greatest success found for detecting the denial of self-referring information such as one's birthdate (e.g., Rosenfeld, Shue, & Singer, 2007; for review see Rosenfeld, 2002).

Neural activity during true and false memories

Although brain electrical activity differentiates the encoding of lures later falsely recalled from those not falsely recalled, showing a widespread topography of differences 500–1300 ms poststimulus during the encoding task (Urbach, Windmann,

Payne, & Kutas, 2005), differences at recognition between falsely recognized and accurately recognized items are rather subtle, if present at all. Johnson et al. (1997) examined ERPs-associated true and false recollections in the DRM paradigm as a function of blocked vs. random intermixed stimulus presentation. Within blocked conditions, true memories evinced greater amplitudes as compared to false memories at more anterior sites (FP1, FP2, F3, F4) in the 50–775 ms range. By contrast, no differences were observed in a randomized stimulus presentation condition. Replicating previous findings (Johnson, Kounios, & Nolde, 1996). Johnson et al. (1997) also found larger ERP amplitudes during correct recognition as compared to false recognition during a later time range (775–1500 ms) in the blocked condition only. They argued that waveform differences between conditions were important since the relative semantic equality and familiarity among stimuli in the blocked condition forced participants to engage in more detailed perceptual-contextual based decision processes; a process not required in the random condition. The random intermixed presentation paradigm is like that used in most of the DRM literature, except for a few imaging studies (e.g., Schacter et al., 1996).

Miller, Baratta, Wynveen, and Rosenfeld (2001) investigated the effects of true and false recognition in terms of P3 amplitude, latency and topography. Employing mass stimulus presentation in the DRM paradigm, no amplitude or topography differences of the P3 across conditions were observed with the limited electrode array, but false recognition of critical lure words were associated with significantly shorter P3 latencies as compared to latencies associated with true recognition of studied words at one electrode site. Miller et al. hypothesized that the observed latency difference "may reflect differences between true and false recognition at the stimulus evaluation level that do not affect response selection" (p. 360).

The present studies

The purpose of the present two studies was to build on the findings of Johnson et al. (1997) and Miller et al. (2001), to specifically determine whether under certain conditions, differences may in fact emerge between veridical and false memories that would be apparent in brain electrical activity, and whether any differentiation would be useful for assessing individual subjects. The present studies employed an intermixed study–test so that participants completed a recognition test immediately after presentation of each study list. By minimizing the time elapsed between initial encoding and subsequent recognition, signal differences between lure and learned words have little time to decay, which should maximize chances to differentiate these conditions. Moreover, the present investigation used a denser recording array than Johnson et al. (1997) and Miller et al. (2001), which might allow for the detection of spatially small effects that would distinguish true from false memories.

The present studies controlled for word frequency effects, as word frequency has been shown to exert a powerful influence on late positive potential amplitude (Polich & Donchin, 1988). Additionally, Johnson et al. (1997) did not exploit the known sensitivity of the ERP to probability effects. It is possible that presenting learned targets and critical lures infrequently, against a backdrop of matched words that had not been learned, would cause differences to become apparent that would be obscured in the equiprobable presentation conditions. By decreasing the probability of the critical and learned items, one would increase the amplitude of the P3 potential, possibly increasing the chance to detect what might be a small difference between learned and critical items. Study 1 employed a dichotomous recognition response format, whereas Study 2 examined confidence ratings. Mather, Henkel, & Johnson (1997) found that confidence levels differentiated true from false memories (i.e., individuals were less confident about false memories as compared to true memories), one of the criteria suggested by Johnson, Hashtroudi, and Lindsay (1993) as helpful for correct memory source monitoring. Confidence ratings were used in study 2 to explore whether engaging in more detailed decision processes—as compared to a simple yes/no response—might reveal additional waveform differences analogous to the differences Mather et al. (1997) were able to show on a paper and pencil evaluation of such memories.

STUDY 1

Methods

Participants

Twenty[1] undergraduate students (5 M, 15 F, mean age = 20.8, *SD* = 2.8) from the University of Arizona were native English speakers, naïve to the task, free of psychological or neurological disorders by self-report, and produced at least 10 (out of a possible 31) behavioral responses indicating false recognition of lure words. Participants provided informed consent and received course credit for their participation.

Materials and procedures

Learning task. Thirty-one 15-item stimulus lists obtained from Roediger and McDermott (1995) and Russell and Jenkins (1952) were used during the learning task. All 15 words within each list were highly associated to a critical lure that was never presented during the learning session. Within stimulus lists, items were presented by strength of association with the critical lure; strongest associates occurred first, weakest occurred last (e.g., for the critical lure *Needle* the list was presented in the following order: thread, pin, eye, sewing, sharp, point, prick, thimble, haystack, thorn, hurt, injection, syringe, cloth, knitting). The 31 lists were presented in one of two counterbalanced orders. Each list was heard via audiotape by a female speaker at a rate of one word every 2 s. No electroencephalographic (EEG) activity was recorded during the presentation of these stimulus lists.

Recognition task. Immediately following the presentation of each stimulus list, participants viewed a 12-item recognition list presented on a computer monitor using DMDX stimulus presentation software (Forster & Forster, 2003)[2] at a rate of one word every 3 s. Each recognition list consisted of one previously heard item from the previous encoding list (learned), the critical lure from that list (which was never presented in the learning task) and ten distractor items (unlearned). Serial position was randomized within each recognition list using the following two constraints: Learned and critical lure (a) did not occur as the first item in the list; and (b) were not presented consecutively. To control for word frequency effects on the P300 (Polich & Donchin, 1988), learned and distractor (unlearned) items were matched to the critical lure by using the Kucera-Francis written word frequency database.[3] The learned item for testing from each list was determined by selecting the word on the list that had the closest word frequency match to the critical lure. Distractor items were selected based on two criteria: word length between three and eight characters, and a word frequency ± 10/million above and below the range of the critical item and learned item. The success of the word frequency matching across lists was evaluated statistically, with no statistically significant difference in word frequency observed between learned, critical lure, and distractor items, $F(2, 60) = 2.46$, *ns*. Each word in the recognition task occurred only once throughout the entire paradigm. EEG activity was recorded during this recognition task.

Instructions were presented auditorily and additionally displayed visually on a monitor. Participants were told that they would be listening to a recording of lists of words and that immediately following each list they would be asked to indicate via button presses whether a word shown on a monitor was a word they had heard in the preceding list. All participants received instructions to press a joystick button with the thumb of their right hand if the word they saw was a word in the previously heard list

[1] The Department of Psychology does not permit the exclusion of study participants based on handedness or native language. To comply with these requirements, 38 undergraduate students expressing interest were enrolled in Study 1; 4 participants were excluded for not being exclusively right-handed and 6 excluded because they were non-native English speakers. One additional student was excluded after acknowledging having been aware of the paradigm after data collection was completed. Data for 5 participants were never analyzed due to equipment problems and excessive artifacts. Data from the remaining 22 students were initially analyzed, but 2 additional participants were excluded from the analyses because they failed to produce at least 10 behavioral responses indicating false recognition of lure words. Inclusion of these 2 participants changed the overall recognition rates to 73% for Lures, 66% for Learned, and 5% for Distractor items.

[2] The display software used for Study 1 was the DOS-based DMTG, and the display software for Study 2 was the Windows based DMDX, both developed at the University of Arizona by J. C. Forster and K. I. Forster. A detailed description of these systems can be found at www.u.arizona.edu/ ~jforster/dmdx.htm

[3] Available at http://www.psy.uwa.edu.au/MRCDataBase/uwa_mrc.htm

and to press a joystick button with the thumb of their left hand if a word they saw was not a word in the previously heard list.

Electrophysiological recording

EEG activity was recorded at 26 scalp sites placed via the 10–20 electrode system by using a Lycra electrode cap: Fp1, Fp2, F7, F3, Fz, F4, F8, FTC1, FTC2, T3, C3, Cz, C4, T4, TCP1, TCP2, T5, P3, Pz, P4, T6, PO1, PO2, O1, Oz, O2. All electrodes were referenced online to Cz with a forehead ground and electrode impedances at 5 kΩ or less. To enable the offline derivation of the customary link–mastoids reference, montage electrodes were also placed on each mastoid (A1, A2) and referenced to Cz. To evaluate possible site differences across hemispheres, the homologous interelectrode impedance range was limited to 1 kΩ or less. Electrooculographic (EOG) activity was monitored by placing tin electrodes at the nasion and directly below FP1 and FP2 (using 20% of the nasion–inion distance). EEG recordings were obtained using a digital Synamps amplifier (Compumedics, Inc., El Paso, TX) with signals amplified by a factor of 500, band-limited from 0.1 to 50 Hz, and digitized continuously at 250 Hz. Files were digitally filtered with a 12.5 Hz, 96 dB per octave low-pass filter and then corrected for ocular artifact (Semlitsch, Anderer, Schuster, & Presslich, 1986), using a bipolar channel derived from the nasion and the site below the left eye; following correction, epochs with deflections exceeding ± 75 μV in any EEG channel were rejected as a subsequent precaution against including ocular or other artifacts, and grand average scalp traces were visually inspected as a further precaution to ensure that ocular activity was not an influence. Files were epoched from -250 ms prestimulus onset to 1750 ms poststimulus, linearly detrended, baseline corrected to a prestimulus baseline of -250 to 0 ms and finally re-referenced offline to averaged mastoids. Epoched segments were then sorted and averaged to extract a grand average for: correctly recognized learned items, falsely recognized lure items, and correctly rejected distractor items. Because previous studies indicated very low rejection rates of lure and learned items and very low rates of false recognition of distractor items, these waveform averages were expected to exhibit poor signal-to-noise ratios, and were not analyzed further.

Data analysis

Analyses of peak amplitudes of ERP waveforms were conducted for P1, N1, P2, N2 and P3 components. Search windows were selected based on visual inspection of the grand average waveform. The maximum positive amplitude in the search window was taken as the amplitude measure for P1 (75–200 ms), P2 (200–300 ms) and P3 (300–800 ms) and the maximum negative amplitude in the search window was taken as the amplitude measure N1 (130–230 ms) and N2 (300–500 ms). Amplitude was measured relative to the prestimulus baseline, and in each case was taken at the latency value associated with the peak at that site.

For each component amplitude and latency, a 3 (Condition: Learned, Lure, Distractor) × 3 (Site: Fz, Cz, Pz) ANOVA was conducted. Due to the study's interest in the P3 as potentially sensitive to differences between true and false memories, additional analyses were only performed on P3 components. These follow-up analyses were performed to specifically compare the lure and learned words, excluding the distractor words. P3 follow-up analyses used a 2 (Condition: Learned, Lure) × 2 (Hemispheres: Left, Right) × 5 (Site within hemisphere from anterior to posterior) ANOVA involving off-midline sites across both hemispheres including prefrontal, frontal, central, parietal, and occipital sites (Fp1, Fp2, F3, F4, C3, C4, P3, P4, 01, 02, respectively) per Johnson et al. (1997). ANOVAs were carried out with the Greenhouse-Geisser correction applied where appropriate. Reported are the uncorrected degrees of freedom, the epsilon value, and the probability level following correction. Newman-Keuls comparisons were employed for post-hoc comparisons. Additionally, exploratory analyses for P3 amplitude were motivated by visual inspection of paired *t*-test maps comparing lure and learned words within specific time windows.

Finally, to assess whether any differences between critical lure and truly learned items could be detected on a subject-by-subject basis, as would be important for field applications in deception detection, a Bayesian classification approach was employed. The approach is described in detail in several other papers (Allen, 2002; Allen et al., 1992; Mertens & Allen, 2008), and uses features of participants' ERP waveforms as indicators of whether an ERP was indicative of recognition or nonrecognition. Allen et al. (1992) adopted this approach to classify ERP waveforms as being in response to learned and unlearned word lists, by combining different features of waveforms at site Pz (e.g., P3 amplitude, 1st, 2nd derivative of ERP waveforms), where each feature was somewhat effective in differentiating familiar from unfamiliar items. (Inclusion of additional sites did not improve overall classification accuracy.) Given the indicators, the approach computes the probability that a given ERP is in response to recognized items, given a pattern of indicators. The computed probability is high if all indicators suggest that the ERP was in response to recognized items, whereas disagreement among indicators lowers this probability. This approach produces for each list of items (in the present study: Lure, Learned, and several distractors) a specific probability that the ERP is indicative of recognition of the items that elicited it, regardless of whether subjects overtly indicate recognition or nonrecognition (Allen et al., 1992). Although bootstrapping approaches also exist for such purposes (e.g., Farwell & Donchin, 1991; Rosenfeld, Soskins, Bosh, & Ryan, 2004), such approaches produce highly similar rates of classification accuracy (Allen & Iacono, 1997; Mertens & Allen, 2008) and will thus not be utilized in this report. Using the Bayesian approach, Allen et al. (1992) correctly classified 94% of learned items as recognized, regardless of whether subjects truthfully indicated recognition or followed instructions to conceal having learned a list of words. Moreover, the approach rarely classified distractors as learned, with only 4% of distractor lists incorrectly classified as recognized.

Results

Behavioral performance

Participants reported false recognition (i.e. having previously seen lure words) 79% of the time. By contrast, they correctly recognized learned words 69% of the time and correctly identified distractor words as not previously seen 95% of the time. These recognition rates differed significantly across the three conditions, $F(2, 38) = 21.85$, $\varepsilon = .76$, $p < .001$. Post-hoc comparisons specifically examining learned and lure words indicated that lure words were reported as recognized at higher rates than learned words ($p < .05$). Response latencies were also compared for correct rejections of distractor words (684 ms), correct recognition of learned words (910 ms), and false recognitions of lure words (821 ms). Response latencies differed significantly, $F(2, 38) = 41.66$, $\varepsilon = .91$, $p < .001$, with distractor items having shorter response latencies compared to lure and learned words ($p < .001$ in both cases). Moreover, lure words had shorter response latencies than learned words ($p < .01$).

Event-related potentials

The grand average ERPs for the target stimuli from midline recording sites in Study 1 are presented in Figure 1. As can be seen, both lure and learned words elicit a prominent and large P3 component compared to distractor words. The numbers of trials comprising these averages were: Lure = 22.1, Learned = 18.9, Unlearned = 224.6.

P3 component. P3 amplitude (see top left panel of Figure 2), as assessed with an omnibus 3 (Condition) × 3 (Site) ANOVA, increased reliably from the frontal to parietal electrodes, $F(2, 38) = 33.11$, $\varepsilon = .62$, $p < .01$, and differed across conditions, $F(2, 38) = 33.74$, $\varepsilon = .98$, $p < .01$; these effects were qualified by a significant Condition × Electrode interaction, $F(4, 76) = 3.23$; $p < .05$. Post-hoc comparisons revealed that at all three midline sites, the lure and learned words did not differ from each other ($p > .10$), while both produced significantly larger P3 amplitudes than distractor words across these sites ($p < .001$). A follow-up analysis of the Condition × Electrode interaction indicated that P3 amplitude increased steadily across all midline sites for the distractor words ($p < .001$). As shown in the top left panel of Figure 2, lure and learned words exhibited a lesser increase from Cz to Pz which was found to be not significant for the learned words ($p > .05$).

P3 latency analysis revealed significant differences between conditions as well, $F(2, 38) = 3.98$; $\varepsilon = .79$, $p < .05$ (see lower left panel of Figure 2). Post-hoc comparison indicated no latency differences between lure and learned words ($p > .10$). Distractor words produced shorter latencies as compared to lure words at site Pz ($p < .01$) and produced shorter latencies as compared to learned words at site Cz and Pz only ($p < .05$, $p < .01$, respectively). No additional significant latency findings were obtained.

Because ERP differences between learned and lure words were of specific interest, further analyses assessing hemispheric amplitude differences between lure and learned words at selected sites was assessed via a 2 (Condition) × 2 (Hemisphere) × 5 (Site within hemisphere) ANOVA.

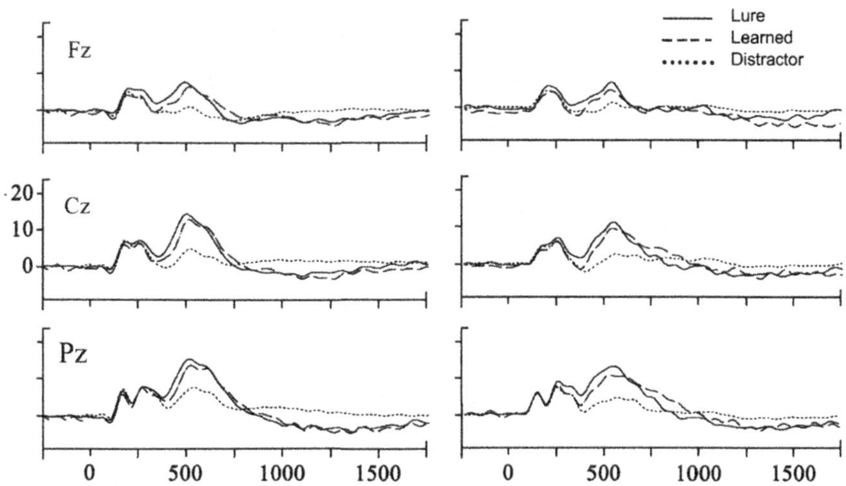

Figure 1. Grand averaged ERP waveforms for Study 1 (left column, $n = 20$) and Study 2 (right column, $n = 23$) depicting falsely recognized Lure, correctly recognized Learned, and correctly classified Distractor items at frontal, central, and parietal midline sites. Both Lure and Learned waveforms include only trials with responses indicating recognition ("yes" for Study 1 and somewhat certain or certain it was heard collapsed for Study 2). Distractor waveforms include only trials with responses indicating nonrecognition ("no" for Study 1 and somewhat certain or certain it was not heard collapsed for Study 2).

Although there was a significant main effect for site, $F(4, 64) = 10.24$, $p < .001$, and a Hemisphere × Site interaction, $F(4, 64) = 5.30$, $p < .001$, no significant effects involving condition emerged.

P1, N1, P2, N2 components. Each component amplitude was assessed with a 3 (Condition) × 3 (Site) ANOVA (see Table 2). Only the P1 component produced a significant Condition × Site interaction, $F(4, 76) = 2.55$, $p < .05$. Subsequent post-hoc comparison indicated that lure words produced significantly larger P1 amplitudes as compared to learned and distractor words at site Fz only ($p < .01$). For the overall findings of these analyses, the reader is referred to Table 2.

Component latency analysis produced a significant condition effect for the N2 component, $F(2, 38) = 8.54$, $\varepsilon = .94$, $p < .001$. Subsequent post-hoc comparison indicated that learned words produced faster N2 latencies ($p < .05$) as compared to lure and distractor words, which in turn did not differ from each other ($p > .05$).

Exploratory analysis

To explore whether any features of the ERP waveform could differentiate falsely recognized lures from correctly recognized learned items, the lure waveform was subtracted from the learned waveform for each subject (retaining variance information), and the grand averaged subtraction

TABLE 2

F-ratios from the 3 (Condition) × 3 (Site) analyses of variance for the P1, N1, P2 and N2 ERP components performed on Amplitude and Latency

	Amplitude (μV)				*Latency (ms)*			
Factor (df)	*P1*	*N1*	*P2*	*N2*	*P1*	*N1*	*P2*	*N2*
Study 1								
Condition (2, 38)	–	–	–	–	–	–	–	8.54***
Site (2, 38)	6.73**	5.61**	14.12***	20.69***	31.50***	15.43***	15.88***	–
C × S (4, 76)	2.55*	–	–	–	–	–	3.92**	
Study 2								
Condition (2, 42)	–	4.71*	6.05**	–	–	3.49*	9.85***	
Site (2, 42)	3.31*	–	6.36**	20.26***	14.45***	25.92***	29.55***	–
C × S (4, 84)	–	–	–	–	–	2.65*	–	–

Notes: *$p < .05$; **$p < .01$; ***$p < .001$.

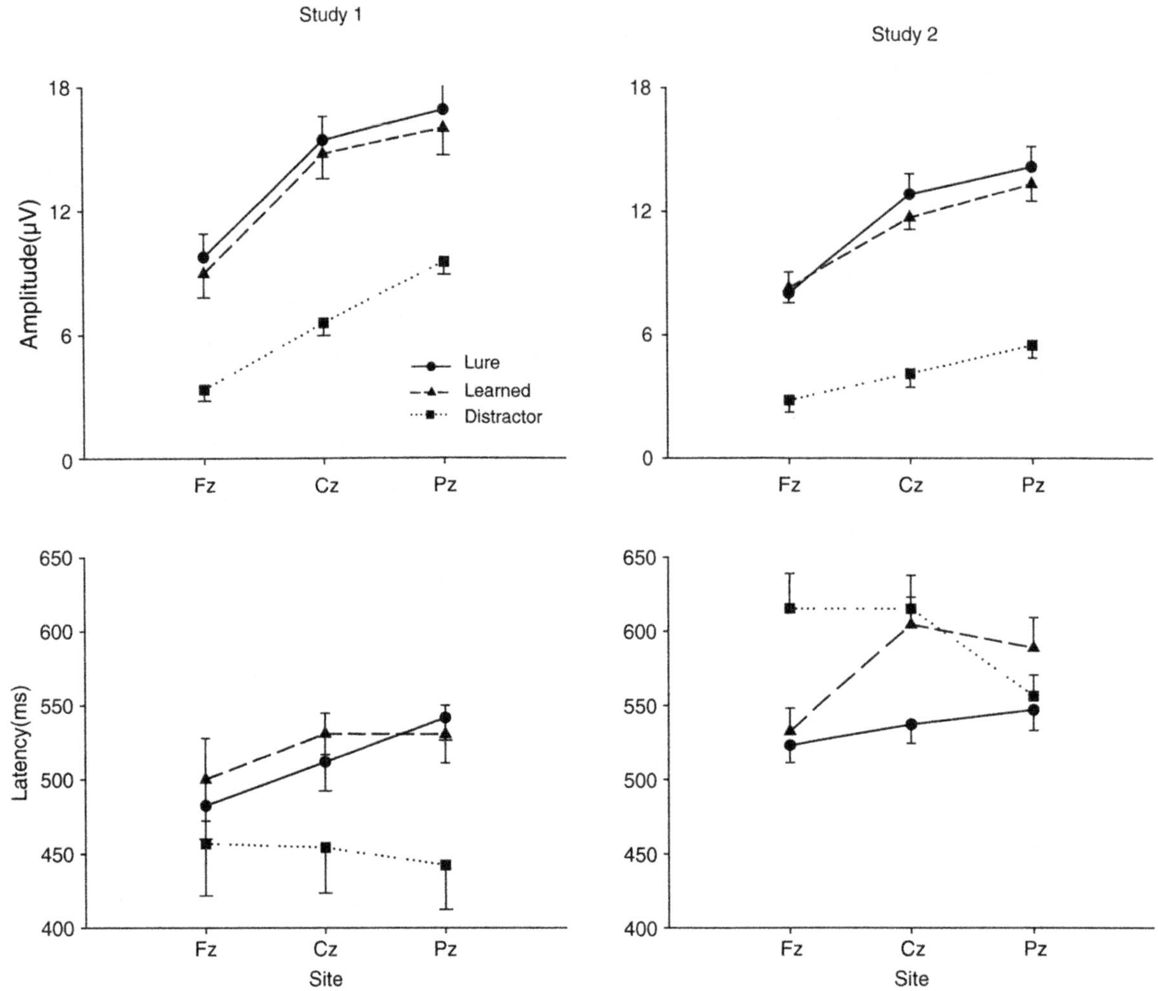

Figure 2. Mean (± *SE*) P3 amplitudes and latencies for all three conditions in Study 1 and Study 2 at frontal, central and parietal midline sites.

waveforms allowed for a time point × time point *t*-test (Duffy, Iter, & Surwillo, 1989). The resulting *t*-test waveforms (for each electrode site) were then mapped on a head model by linearly interpolating four adjacent electrode sites, and averaged over 100 ms bins for the entire 0–1400 ms epoch. *T*-tests at $p < .01$ and $p < .05$ levels of significance did not indicate any significant amplitude differences between lure and learned words at any site in any time window.

Individual classification

Utilizing the Bayesian classification approach described above (and detailed in Allen, 2002; Allen et al., 1992; and Mertens & Allen, 2008), features of participants' ERP waveforms were used as indicators of whether an ERP was in response to recognized or nonrecognized items. Using this approach, learned items were correctly classified as recognized for 65% of the subjects, and distractor items were correctly classified as unrecognized 96% of the time. Critical lures, however, were classified as recognized 80% of the time. Compared to previous work with this approach, the hit rate for learned items is somewhat lower (compared to 94% in Allen et al., 1992), perhaps because learned items in Allen et al. were presented on multiple iterations. On the other hand, the false positive rate for classifying unfamiliar words as recognized is virtually identical.

Because the Bayesian approach depends on a set of cutpoints established in previous work, it may be more informative to compare differentiation of learned from distractor items, and lures

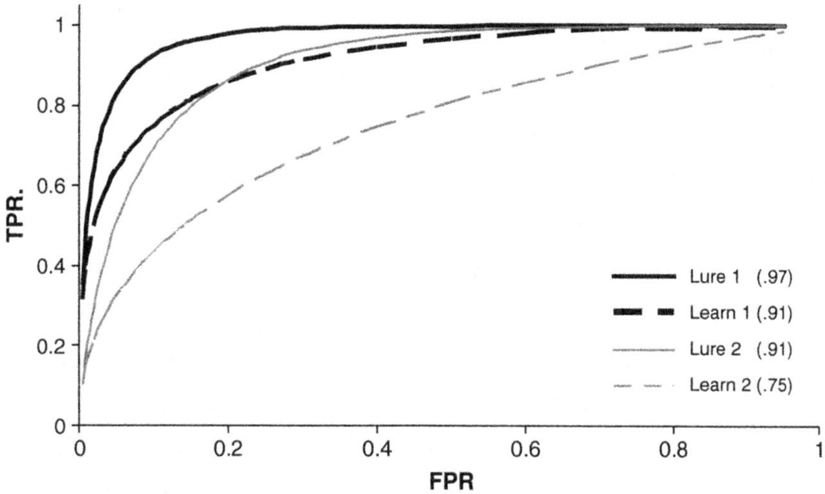

Figure 3. ROC curves illustrating the differentiation of truly learned from distractor items in Study 1 (Learn 1) and Study 2 (Learn 2), and illustrating the differentiation of critical lures from distractor items in Study 1 (Lure 1) and Study 2 (Lure 2). Values in parentheses are AUCs. In Study 1, AUC did not significantly differ ($p > .10$) for learned items and critical lures. In Study 2, critical lures were significantly better differentiated from distractor items than were truly learned items ($p < .01$).

from distractor items, in terms of receiver operating characteristic (ROC) curves. As shown in Figure 3, which compares learned items to distractor items and also compares lure words to distractor items, both learned and lure items have an area under the curve (AUC) of over .90, confirming that the ERP signatures to both lure and learned items are clearly differentiable from those to distractor items, even on the individual subject level. Unfortunately, however, the AUCs for the learned and lure items did not different significantly ($p > .10$), confirming that ERPs are not useful in distinguishing learned from lure items, for any individual subject, limiting their utility in distinguishing true from false recollections.

STUDY 2

Study 1 attempted to maximize the possibility of finding electrocortical differences between learned and lure words by adjusting stimulus probability that maximized P3 amplitude. This manipulation failed to uncover significant differences as manifested in the ERP. Study 2 thus altered response format to a confidence rating, which has been found to create some differentiation between learned and lure items under some conditions (Henson, Rugg, Shallice, & Dolan, 2000; Johnson, Hashtroudi, & Lindsay, 1993). Study 2 was identical to Study 1 except that participants were required to respond with a confidence rating instead of a dichotomous YES/NO judgment.

Methods

Participants

Study 2 had a total of 23 participants[4] (7 M, 16 F, mean age = 20.13, *SD* = 2.28). Selection criteria were identical to Study 1.

Materials and procedures

Materials and procedures were identical to those used in Study 1 with the following exception. Instructions were extended by asking the participants to respond by assigning confidence ratings to each stimulus presentation. Four letters on the keyboard, two to be pressed by the right hand (K, J) and two to be pressed by the left hand (D, F), were designated for confidence ratings, and instructions were presented along with a graphic display demonstrating which key corresponded to which rating. All participants received

[4] Study 2 enrolled a total of 35 undergraduate students. Data from 4 participants were never collected due to equipment problems, 1 student was aware of the paradigm and data from 5 participants showed excessive EEG artifact. Data from the remaining 25 participants was initially analyzed, but 3 additional participants were excluded from the analyses because they failed to produce at least 10 behavioral responses indicating false recognition of lure words. Inclusion of these three participants changed the overall recognition rates to 69% for Lures, 69% for Learned and 7% for Distractor items.

the same instructions: "Using your right middle finger, press K if you are certain that the word you saw is a word you heard. Using your right index finger, press J if you are somewhat certain that the word you saw is a word you heard." Participants received similar instructions for non-recognized items: "Using your left middle finger press D if you are certain that the word you saw is not a word you heard. Using your left index finger press F if you are somewhat certain that the word you saw is not a word you heard." Participants practiced making these ratings until it was clear they understood the rating task.

Electrophysiological recording

Electrophysiological recording and data preparation were identical to those in Study 1, except that averages were created for lure and learned items where participants were either certain or somewhat certain the item was recognized, and for distractor items where participants were either certain or somewhat certain that the item was not previously heard. The distribution of responses and the limited number of trials did not allow for a separate analysis comparing confident to somewhat confident responses; thus data presented reflect the combined averages of these conditions.

RESULTS

Behavioral performance

Participants expressed high confidence in their decision with very low endorsement of the "somewhat confident" category. Separate analysis comparing confident to somewhat confident responses thus were not possible. The following data reflect the combined averages of these conditions. Participants recognized (confident or somewhat confident) learned words 69% of the time, falsely recognized lure words previously 76% of the time, and correctly identified distractor words as not previously seen 95% of the time, producing a significant effect across condition, $F(2, 44) = 24.86$, $\varepsilon = .94$, $p < .001$.

As the four categorical ratings of confidence were collapsed into two categories because some cells had very low endorsement rates (i.e., participants being somewhat confident of having heard or not heard an item), subtle differences in confidence may have been obscured. A continuous measure was thus derived to reflect the confidence with which participants rated each word as being familiar. For each subject and for each condition, a weighted average was constructed such that confidently heard responses were given a weight of 4, somewhat confidently heard were weighted 3, somewhat confidently not heard were weighted 2, and confidently not heard were weighted 1. The number of responses for each subject in each condition was multiplied by the corresponding weight, summed, and divided by the total number of responses. Scores on this metric could range from 1, in cases where all items were rated as confidently not heard, to 4, in cases where all items were rated as confidently heard.

A clear differentiation, $F(2, 44) = 260.79$, $\varepsilon = .91$, $p < .001$, between word types emerged, with participants having significantly higher confidence of recognition for lure and learned words (3.22 and 3.00, respectively) as compared to distractor words (1.11, high confidence of non-recognition). Post-hoc tests indicated that confidence for lure words was greater than that for learned words ($p < .05$), and that distractor words differed significantly from lure and learned words ($p < .001$, both cases).

Event-related potentials

The grand average ERPs for the target stimuli from midline recording sites in Study 2 are presented in Figure 1. As can be seen from the figure, both lure and learned words elicit a prominent and large P3 component compared to distractor words. The number of trials comprising these averages were: Lure = 19.0, Learned = 19.5, Unlearned = 215.0.

P3 component, confidence collapsed. As in Study 1, and depicted in the top right panel of Figure 2, P3 amplitude increased reliably from the frontal to parietal sites, $F(2, 42) = 30.63$, $\varepsilon = 63$, $p < .001$, differed across conditions, $F(2, 42) = 66.10$, $\varepsilon = .83$, $p < .001$, and was qualified by a significant Condition × Electrode interaction, $F(4, 84) = 8.74$, $p < .001$. Post-hoc tests revealed that at all three midline sites, both the lure and learned words produced significantly larger P3 amplitudes than distractor words ($p < .001$). In contrast to Study 1, however, lure words produced significantly larger P3 amplitudes than did learned words at Cz and Pz, ($p < .05$, $p < .01$ respectively). To assess whether the P3 difference may have been the result of greater trial-to-trial latency jitter for learned than lure words, waveforms were reanalyzed using latency-shifted waveforms (Woody, 1967); the amplitude differences between lure and learned words remained.[5]

P3 latency analyses (see bottom right panel in Figure 2) revealed significant differences between condition, $F(2, 42) = 5.56$, $\varepsilon = .80$, $p < .05$, but not electrode site, $F(2, 42) = 2.70$, $p > .05$. A significant Condition × Site interaction, $F(4, 84) = 7.88$, $p < .001$, was obtained. Post-hoc evaluation identified significantly shorter P3 latency of lure words compared to learned words at sites Cz and Pz ($p < .01$, $p < .05$, respectively), whereas P3 latency was significantly shorter for lure words as compared to distractor words at sites Fz and Cz ($p < .001$ in both cases). No other significant latency effects were observed.

Because ERP differences between learned and lure words were of specific interest, further analyses assessing hemispheric amplitude differences between lure and learned words at selected sites were assessed via a 2 (Condition) × 2 (Hemisphere) × 5 (Site within hemisphere) ANOVA. Results were similar to those observed in Study 1. P3 amplitude varied with site, $F(4, 80) = 18.22$, $p < .001$, and demonstrated a significant Site × Hemisphere interaction, $F(4, 80) = 3.46$, $p < .001$, but no significant effects involving condition emerged. In general, right hemisphere amplitudes were larger as compared to the left hemisphere across prefrontal to central sites ($p < .05$), with the exception of the parietal sites, at which left hemisphere amplitudes were larger than right hemisphere amplitudes ($p < .05$). No amplitude differences were observed across occipital sites ($p > .05$).

[5] To assess whether latency jitter across trials may have contributed to the amplitude differences, the approach described by Woody (1967) was used to align single trials with predefined sine-wave template that was centered on 550 ms. The procedure shifted trials up to ± 125 ms, selecting the shift that maximally correlated with the template. Subsequent iterations then used the average of latency-shifted trials from the previous iteration as a template, and again shifted epochs to maximize the correlation with the template. Iterations terminated when the average correlation of single trials with the template increased less than .005 from the previous iteration. The output of this procedure then allowed for an examination of the degree of latency shift required to fit individual trials to the grand average. Analysis of these latency shifts revealed no significant difference across lure and learned conditions in terms of either mean latency shift, $F(2, 44) = 0.31$, *ns*, or the absolute value of these latency shifts, $F(2, 44) = 0.95$, *ns*, suggesting that the extent and direction of latency jitter were not different between these conditions.

P1, N1, P2 and N2 component. Amplitudes and latencies were tested in a 3 (Condition) × 3 (Site) ANOVA. Investigation of the N2 amplitude component revealed a condition main effect, $F(2, 42) = 6.05$, $p < .01$, with less negative amplitudes for the lure words as compared to the learned and distractor words ($p < .05$, $p < .01$, respectively) whereas the two latter components did not differ from each other ($p > .10$). Latency data revealed significant condition main effects for P2 and N2, $F(2, 42) = 3.49$, $p < .05$; $F(2, 42) = 9.85$, $p < .001$, respectively. P2 latencies were due to faster latencies for lure words as compared to learned words ($p < .05$). Post-hoc comparisons performed on the condition main effect of the N2 indicate longest latencies for the distractor words as compared to lure and learned words ($p < .01$), with no difference between the lure and learned condition ($p > .10$). No other significant latency findings were observed.

Exploratory analyses

A visual inspection of the midline ERP plots depicted in Figure 4 suggests a different recovery rate of P3 amplitude to baseline between lure and learned words during a later phase of the epoch (800–1000 ms) at frontal, central and parietal sites adjacent to the midline. To explore possible differences, *t*-test maps were created in the same way as described in Study 1, with the exception that a longer epoch was investigated (0.0–1750 ms). The *t*-test map depicted in Figure 4 indicated a larger area under the curve for the lure as compared to the learned words ($p < .05$, $T > 2.074$, $df = 22$) in the central and parietal region (400–600 ms) as previously reported in the omnibus ANOVA. Additional, significant differences between lure and learned words were found during the recovery of the lure and learned words to baseline (800–1000 ms), with learned words recovering more slowly as compared to the lure words ($p < .05$). To statistically evaluate the difference in rate of recovery between lure and learned words, a 2 (Condition: Learned, Lure) × 2 (Hemisphere: Left, Right) × 3 (Site: Frontal, Central, Parietal) ANOVA was conducted on the mean amplitude during the interval from 800 to 1000 ms at off-midline sites (F3, F4, C3, C4, P3, P4). Area analysis indicated main effects across condition, $F(1, 22) = 7.99$, $p < .01$, and hemisphere, $F(1, 22) = 22.35$; $p < .01$. Moreover, a Condition × Site interaction, $F(2, 44) = 4.19$; $p < .05$, and a Hemisphere × Site, $F(2, 44) = 5.28$, $p < .01$, effect were observed.

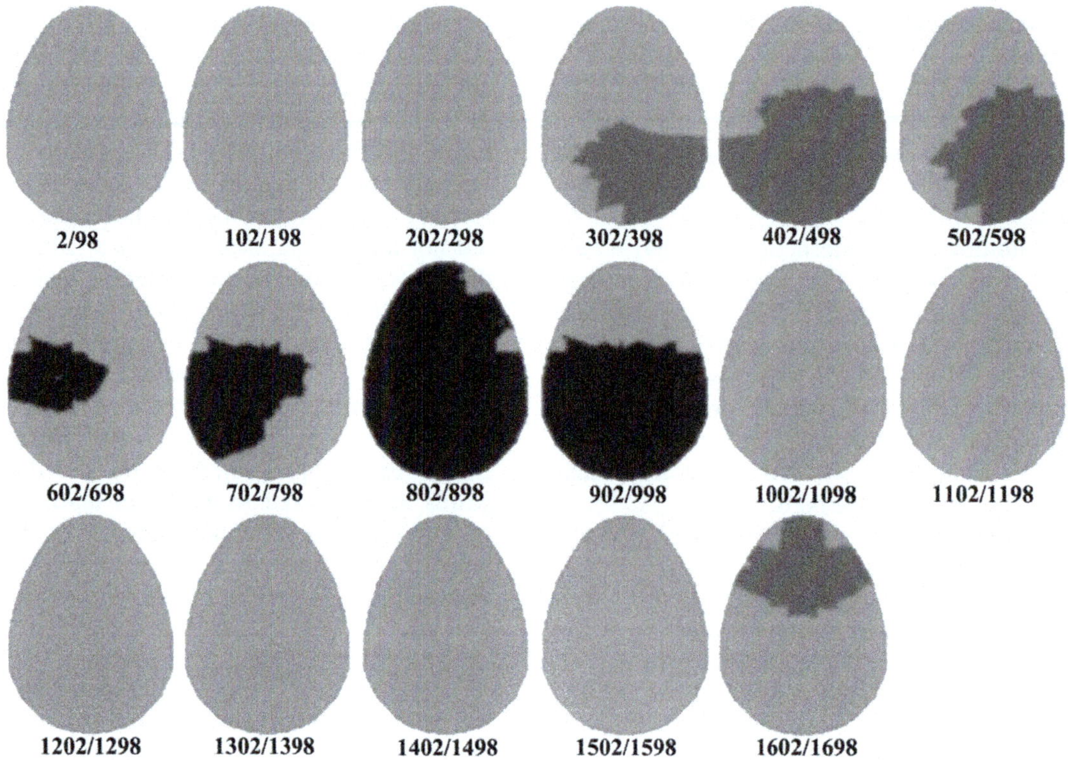

Figure 4. The exploratory *t*-score comparison of lures minus learned words at the $p < .05$ level ($T > 2.074$, $df = 22$). Colors other than light gray indicate significant differences. Dark gray indicates larger positive amplitudes in response to lure words as compared to learned words, whereas black indicates larger positive amplitudes to learned words as compared to lure words.

Learned words were associated with larger amplitude areas ($p < .01$) as compared to lure words, while the right hemisphere produced larger values than the left across both conditions ($p < .001$). Area values were consistently larger in the right hemisphere across frontal, central, and parietal regions ($p < .01$).

Because differences between lures and learned items were observed under conditions of confidence rating in Study 2, but not under the forced choice conditions of Study 1, the results of the two studies were directly compared in a 2 (Study: 1 and 2) × 2 (Condition: Lure, Learned) × 3 (Site: Fz, Cz, Pz) ANOVA. The condition main effect was significant when pooling the studies, $F(1, 38) = 11.08$, $p < .01$, respectively, but there was no significant interaction of study and condition, $F(1, 38) = 1.53$, *ns*. Examination of the effects sizes for the list effect in separate List (Lure, Learned) × Site (Fz, Cz, Pz) ANOVAs for each study reveals that the effect size (η^2) for differentiating lure from learned lists was descriptively, but not statistically, larger in Study 2 ($\eta^2 = .29$) than in Study 1 ($\eta^2 = .15$).

Individual classification

The Bayesian classification approach was again applied to ascertain how well ERPs could distinguish learned from lure items for individual subjects. Using this approach, learned items were correctly classified as recognized for 43% of the subjects, and distractor items were correctly classified as unrecognized 95% of the time. Critical lures were classified as recognized 61% of the time. ROC curves were again computed to characterize the differences between learned and lure items for individual subjects. As shown in Figure 3, which compares learned items to distractor items and also compares lure words to distractor items, lure items have a significantly ($p < .01$) larger AUC (.91) compared to learned items (.75). Thus, although learned and lure words were significantly different, the AUC for the learned items was in fact smaller than that for the lure items, suggesting that lure items have more features suggestive of recognition that truly learned words. Thus again, as in study 1, these results suggest that brain electrical activity to falsely recognized items is clearly distinct from that in response to unrecognized items. ERPs therefore would have little or no use in effectively differentiating false from true recognition, as any recognized item will elicit brain activity quite different than that to items that in fact were not previously experienced.

DISCUSSION

Controlling for word frequency, and exploiting the known sensitivity of the P3 component to subjective probability of occurrence, the present studies found limited evidence for differentiation between valid and false recollections. Both studies elicited high rates of false recognition, similar to those reported by other studies utilizing the DRM paradigm (e.g., Johnson et al., 1997; Roediger & McDermott, 1995). Electrocortical differentiation was limited, achieving statistical significance only in Study 2 using a confidence rating procedure. P3 amplitude in response to lure words was in fact larger and P3 latency was faster compared to learned words in that study. Thus, in terms of the indicators of recognition, falsely recognized lures appear to have as many if not more indications that the items are recognized, suggesting that ERP-based GKT procedures may have limitations in field settings where false recollections may be present.

These results add to the small number of ERP studies utilizing the P3 that found latency and amplitude differences between lure and learned words (Miller et al., 2001, and McGill, 2001, respectively). An additional differentiation not reported by other investigators was found during an exploratory analysis assessing the differential return of P3 amplitude to baseline in the 800–1000 ms range, with learned words characterized by a slower return to baseline than lure words. Although this difference was only significant in Study 2, direct comparison to Study 1 revealed that the two studies did not statistically differ in this effect, but merely that the descriptively smaller effect size for this late positivity in Study 1 failed to achieve statistical significance. Thus is not possible to conclude that the confidence manipulation was responsible for finding differentiation in Study 2 but not Study 1.

Study 2 also found shorter P3 latencies for lured words compared to learned words, consistent with the interpretation that lure words appear to be an easier stimulus to be processed. P3 latency "appears to index the mental chronometry of stimulus discriminations" (Polich, Howard, & Starr 1983, p. 668), with P3 latency providing an index of the processing time required prior to response generation (Polich & Herbst, 2000). Lure words may be easier to process because these words have been semantically primed by the list of associates, an inherent feature of the DRM task. McDermott (1997) found limited evidence for such semantic priming effect in the DRM task. McKone and Murphy (2000) found evidence for substantial priming effects when study and test modality were the same, but concluded that this priming is modality-specific, as even though semantic priming exists under certain conditions, it is substantially reduced under crossmodal study–test conditions like those used in both the present studies.

Although the crossmodal design thus reduces the semantic priming effects, it remains possible that the observed ERP differentiation of lure from learned words may simply reflect semantic priming effects that would be expected to be stronger with lure than learned words; several lines of evidence mitigate this possibility. First, the learned item has an inherent semantic relatedness to the list items, as all items are related to the lure. One might predict that the lure is thus more strongly primed than a learned word, but it is nonetheless primed. To the extent that semantic priming may underlie the ERP differentiation of learned and lure items, it should do so in the N2/N4 region, the region shown to be sensitive to semantic priming effects (Deacon, Mehta, Tinsley, & Nousak, 1995; Kutas & Hillyard, 1980; Kutas & Van Petten, 1987). Such effects were not found in the present studies. Moreover, in several studies of ERPs in semantic priming (Koivisto & Revonsuo, 2001; Kutas & Hillyard, 1989; Weisbrod et al., 1999), results indicate that the semantically primed word shows greater positivity in the late time window, whereas the lure shows less positivity than the learned work in this window in the present studies.

The smaller P3 amplitudes for the learned words in Study 2 also appear to support the hypothesis that lure words are more easily classified as compared to learned words. Cutmore and Muckert (1998) suggested that smaller P3 amplitudes may be attributed to processing a stimulus that involves a difficult discrimination in which subjects may have lower confidence. This observation is supported by behavioral data indicating lower confidence when recognizing learned words as compared to lure words. The differential return of P3 amplitude to baseline for learned and lure words is consistent with the need for additional processing for learned words compared to lures. Other ERP studies utilizing the DRM task (cf. Johnson et al., 1997; Mecklinger, 2000; Miller et al., 2001,) show a similar effect in the waveforms, although these investigators did not specifically analyze the time region to ascertain differential responding of lure and learned words. This late differentiation observed in the present Study 2 may in part reflect slow wave (Squires, Squires, and Hillyard, 1975). Slow wave has been shown to increase as confidence in judgments decreases (Kerkhof, 1982), to increase as accuracy decreases (Ruchkin, Sutton, Kietzman, & Silver, 1980), and may be more generally related to the amount of additional processing or mobilization of effort required following the presentation of a stimulus (Friedman, Vaughan, & Erlenmeyer-Kimling, 1981; Ruchkin et al., 1980; Sutton & Ruchkin, 1984). Although lure items provide a strong sense of confidence and familiarity, participants confronted with the learned items may have resorted to what Johnson et al. (1997, p. 256) described as a strategy to "assess perceptual and contextual qualities of their memories". This type of decision process would take more time and cognitive resources due to more exhaustive computational requirements which in turn may be reflected in slow wave, which was significant across both studies combined, with a larger effect size in Study 2.

Various accounts of false recognition memory have been proposed, and two are specifically notable with respect to the present interpretations. Source monitoring (Johnson et al., 1993) provides an explanation on how people identify and differentiate memories from different sources by utilizing semantic, perceptual, and affective content of past situations. False recollection occurs if individuals mistakenly classify an incorrect source as the origin of their memory. Another account of false memories that has received increased attention over the past few years is fuzzy trace theory (Reyna & Brainerd, 1995). This approach postulates that memories are encoded in two parallel, yet independent streams: Verbatim memories refer to perceptual aspects of a particular event such as color, noise, time, whereas gist traces refer to more abstract aspects of the same event such as meaning or interpretations of the event. According to fuzzy trace theory, false memories occur when individuals rely solely on gist information, when gist information is so strong that it is mistaken for verbatim information, or when verbatim traces from another event are mistaken as stemming from the one under consideration. The present data are consistent with both perspectives, in that lure items may rely especially heavily on gist processing, whereas learned items may prompt somewhat greater contextual analysis and thus require further processing reflected in the late positivity observed in the present investigation.

The present studies were not specifically designed to compare these theories' explanatory power, but rather to determine whether it would be possible to differentiate veridical from false memory traces. Consistent with most of the literature, differences between true and false recollections are not often found, and those that are uncovered are relatively subtle, and unlikely to be of use in classifying outcomes for individual subjects, as would be required in deception detection applications. Perhaps future efforts might yet uncover larger and more robust effects, but unless such effects emerge, the difficulty in assessing the veracity of memory in forensic settings and clinical settings will remain.

REFERENCES

Adler, K. (2007). *The lie detectors: The history of an American obsession.* New York: Simon & Schuster.

Allen, J. J. B (2002). The role of psychophysiology in clinical assessment: ERPs in the evaluation of memory. *Psychophysiology, 39*, 261–280.

Allen, J. J. B. & Iacono, W. G. (1997). A comparison of methods for the analysis of event-related potentials in deception-detection. *Psychophysiology, 34*, 234–240.

Allen, J. J., Iacono, W. G., & Danielson, K. D. (1992). The identification of concealed memories using the event-related potential and implicit behavioral measures: A methodology for prediction in the face of individual differences. *Psychophysiology, 29*(5), 504–522.

Bartlett, F. C. (1932). *Remembering: A study in experimental and social psychology.* Cambridge, UK: Cambridge University Press.

Bashore, T. R., & Rapp, P. E. (1993). Are there alternatives to traditional polygraph procedures? *Psychological Bulletin, 113*, 3–22.

Brainerd, C. J., & Reyna, V. F. (1998). When things that were never experienced are easier to "remember" than things that were. *Psychological Science, 9*(6), 484–489.

Brainerd, C. J., Wright, R., Reyna, V. F, & Mojardin, A. H. (2001). Conjoint recognition and phantom recollection. *Journal of Experimental Psychology: Learning, Memory, & Cognition, 27*(2), 307–327.

Budson, A. E., Daffner, K. R., Desikan, R., & Schacter, D. L. (2000). When false recognition is unopposed by true recognition: Gist-based memory distortion in Alzheimer's disease. *Neuropsychology, 14*(2), 277–287.

Cutmore, T. R. H., & Muckert, T. D. (1998). Event-related potentials can reveal differences between two decision-making groups. *Biological Psychology, 47*, 159–179.

Deacon, D., Mehta, A., Tinsley, C., & Nousak, J. M. (1995). Variation in the latencies and amplitudes of N400 and NA as a function of semantic priming. *Psychophysiology, 32*(6), 560–570.

Deese, J. (1959). On the prediction of occurrence of particular verbal intrusions in immediate recall. *Journal of Experimental Psychology, 58*(1), 17–22.

Donchin, E., & Coles, M. G. (1988). Is the P300 component a manifestation of context updating? *Behavioral and Brain Sciences, 11*, 357–427.

Duffy, F. H. P., Iter, V. G., & Surwillo, W. W. (1989). *Clinical electroencephalography and topographic brain mapping: Technology and practice.* New York: Springer-Verlag.

Ekman, P. & O'Sullivan, M. (1991). Who can catch a liar? *American Psychologist, 46*, 913–920.

Fabiani, M., Stadler, M. A., and Wessels, P. M. (2000). True memories but not false ones produce a sensory signature in human lateralized brain potentials. *Journal of Cognitive Neuroscience, 12*, 941–949.

Farwell, L. A., & Donchin, E. (1991). The truth will out: Interrogative polygraphy ('lie detection') with event-related brain potentials. *Psychophysiology, 28*, 531–547.

Farwell, L. A., & Smith, S. S. (2001). Using brain MERMER testing to detect knowledge despite efforts to conceal. *Journal of Forensic Sciences, 46*, 135–143.

Forster, K. I., & Forster, J. C. (2003). DMDX: A Windows display program with millisecond accuracy. *Behavior Research Methods, Instruments, and Computers, 35*, 116–124.

Friedman, D., & Johnson, R. (2000). Event-related potential (ERP) studies of memory encoding and retrieval: A selective review. *Microscopy Research and Technique, 51*, 6–28.

Friedman, D., Vaughan, H. G., & Erlenmeyer-Kimling, L. (1981). Multiple late positive potentials in two visual discrimination tasks. *Psychophysiology, 18*, 635–649.

Gallo, D. A., McDermott, K. B., Percer, J. M., & Roediger, H. L., III. (2001). Modality effects in false recall and false recognition. *Journal of Experimental Psychology: Learning, Memory and Cognition, 27*(2), 339–353.

Ganis, G., Kosslyn, S. M., Stose, S., Thompson, W. L., & Yurgelun-Todd, D. A. (2003). Neural correlates of different types of deception: An fMRI investigation. *Cerebral Cortex, 13*, 830–836.

Garry, M., Manning, C., Loftus, E., & Sherman, S. J. (1996). Imagination inflation: Imagining a childhood event inflates confidence that it occurred. *Psychonomic Bulletin & Review, 3*(2), 208–214.

Henson, R. N. A., Rugg, M. D., Shallice, T., & Dolan, R. J. (2000). Confidence in recognition memory for words: Dissociating right prefrontal roles in episodic retrieval. *Journal of Cognitive Neuroscience, 12*(2), 912–923.

Hicks, J. L., & Marsh, R. L. (2001). False recognition occurs more frequently during source identification than during old–new recognition. *Journal of Experimental Psychology: Learning, Memory and Cognition, 27*(2), 375–383.

Iacono, W. G. (2000). The detection of deception. In J. T. Cacioppo & L. G. Tassinary (Eds.), *Handbook of psychophysiology* (2nd ed., pp. 772–793). New York: Cambridge University Press.

Iacono, W. G., & Lykken, D. T. (1997). The validity of the lie detector: Two surveys of scientific opinion. *Journal of Applied Psychology, 82*, 426–433.

Iacono, W. G., & Patrick, C. J. (1997). Polygraphy and integrity testing. In R. Rogers (Ed.), *Clinical assessment of malingering and deception* (2nd ed., pp. 252–281): New York: Guilford Press.

Iacono, W. G., & Patrick, C. J. (2006). Polygraph ("lie detector") testing: Current status and emerging trends. In I. B. Weiner & A. K. Hess (Eds.), *The handbook of forensic psychology* (3rd ed., pp. 552–588). Hoboken, NJ: Wiley.

Johnson, M. K., Hashtroudi, S., & Lindsay, D. S. (1993). Source monitoring. *Psychological Bulletin, 114*(1), 3–28.

Johnson, M. K., Kounios, J., & Nolde, S. F. (1996). Electrophysiological brain activity and memory source monitoring. *NeuroReport, 7*, 2929–2932.

Johnson, M. K., Nolde, S. F., Mather, M., Kounios, J., Schacter, D., & Curran, T. (1997). The similarity of brain activity associated with true and false recognition memory depends on test format. *Psychological Science, 8*(3), 250–257.

Kerkhof, G. A. (1982). Event-related potentials and auditory signal detection: Their diurnal variation in morning-type and evening-type subjects. *Psychophysiology, 19*, 94–103.

Koivisto, M., & Revonsuo, A. (2001). Cognitive representations underlying the N400 priming effect. *Cognitive Brain Research, 12*, 487–490.

Kozel, F. A., Johnson, K. A., Mu, Q., Grenesko, E. L., Laken, S. J., & George, M. S. (2005). Detecting deception using functional magnetic resonance imaging. *Biological Psychiatry, 5*(8), 605–613.

Kutas, M., & Hillyard, S. A. (1980). Reading senseless sentences: Brain potentials reflect semantic incongruity. *Science, 207*, 203–205.

Kutas, M., & Hillyard, S. A. (1989). An electrophysiological probe of incidential semantic association. *Journal of Cognitive Neuroscience, 1*, 38–49.

Kutas, M., & Van Petten, C. K. (1987). Psycholinguistics electrified. Event-related brain potential investigation. In M. A. Gernsbacher (Ed.), *Handbook of psycholinguistics* (pp. 83–143). New York: Academic Press.

Langleben, D. D., Loughead, J. W., Bilker, W. B., Ruparel, K., Childress, A. R., Busch, S. I., et al. (2005). Telling truth from lie in individual subjects with fast event-related fMRI. *Human Brain Mapping, 26*(4), 262–272.

Loftus, E. F. (1981). *Memory and its distortions.* The G. Stanley Hall Lecture Series. Washington, DC: American Psychological Association.

Loftus, E. F. (1986). Ten years in the life of an expert witness. *Law and Human Behavior, 10*, 241–263.

Loftus, E. F. (1993). The reality of repressed memories. *American Psychologist, 48*(5), 518–537.

Loftus, E. F., Coan, J. A., & Pickrell, J. E. (1996). Manufacturing false memories using bits of reality. In L. Reder (Ed.), *Implicit Memory and Metacognition* (pp. 195–227). Hillsdale, NJ: Lawrence Erlbaum Associates.

Loftus, E. F., & Palmer, J. C. (1974). Reconstruction of automobile destruction: An example of the interaction between language and memory. *Journal of Verbal Learning and Verbal Behavior, 13*(5), 585–589.

Loftus, E. F., & Pickrell, J. E. (1995). The formation of false memories. *Psychiatric Annals, 25*(12), 720–725.

Lykken, D. T. (1959). The GSR in the detection of guilt. *Journal of Applied Psychology, 43*, 385–388.

Mather, M., Henkel, L.A., & Johnson, M. K. (1997). Evaluating characteristics of false memories: Remember/know judgments and memory characteristics questionnaire compared. *Memory and Cognition, 25*(6), 826–837.

McDermott, K. B. (1996). The persistence of false memory in list recall. *Journal of Memory and Language, 35*(2), 212–230.

McDermott, K. B. (1997). Priming on perceptual implicit memory tests can be achieved through presentation of associates. *Psychonomic Bulletin and Review, 4*(4), 582–586.

McDermott, K. B., & Roediger, H. L., III. (1998). Attempting to avoid illusory memories: Robust false recognition of associates persists under conditions of explicit warnings and immediate testing. *Journal of Memory and Language, 39*(3), 508–520.

McGill, K. (2001). Using electrophysiological traces to differentiate verbatim memories and inferences from a list of words. *Dissertation Abstracts International: Section B: The Sciences & Engineering, 62*(2B), 721.

McKone, E., & Murphy, B. (2000). Implicit false memory: Effects of modality and multiple study presentations on long-lived semantic priming. *Journal of Memory and Language, 43*, 89–109.

Mecklinger, A (2000). Interfacing mind and brain: A neurocognitive model of recognition memory. Psychophysiology, *37*, 565–582.

Mertens, R., & Allen, J.J.B. (2008). The role of psychophysiology in forensic assessments: Deception detection, ERPs and virtual reality mock crime scenarios. *Psychophysiolgy, 45*, 286–298.

Mertens, R., & Polich, J. (1997). P300 from a single-stimulus paradigm: passive versus active tasks and stimulus modality. *Electroencephalography and clinical Neurophysiology, 104*, 488–497.

Miller, A. R., Baratta, C., Wynveen, C., & Rosenfeld, J. P. (2001). P300 latency, but not amplitude or topography, distinguished between true and false recognition. *Journal of Experimental Psychology: Learning, Memory and Cognition, 27*(2), 354–361.

National Research Counsel. (2003). *The polygraph and lie detection.* Washington, DC: National Academies Press.

Neisser, U. (1966). *Cognitive psychology.* New York: Appleton-Century-Crofts.

Norman, K. A., & Schacter, D. L. (1997). False recognition in younger and older adults: Exploring the characteristics of illusory memories. *Memory and Cognition, 25*(6), 838–848.

Office of Technology Assessment. (1983). *Scientific validity of polygraph testing: A research review and evaluation—A technical memorandum* (Publication OTA-TM-H-15). Washington, DC: Library of Congress.

O'Sullivan, M., & Ekman, P. (2004). The wizards of deception detection. In P. A. Granhag & L. Stromwell (Eds.), *The detection of deception in forensic contexts* (pp. 269–286). Cambridge, UK: Cambridge University Press.

Paller, K. A, Kutas, M., & McIsaac, H. K. (1995). Monitoring conscious recollection via the electrical activity of the brain. *Psychological Science, 6*(2), 107–111.

Payne, D. G., Elie, C. J., Blackwell, J. M., & Neuschatz, J. S. (1996). Memory illusions: Recalling, recognizing, and recollecting events that never occurred. *Journal of Memory and Language, 35*, 261–285.

Payne, J. D., Nadel, L., Allen, J. J. B., Thomas, K. G. F., & Jacobs, W. J. (2002). The effects of experimentally induced stress on false recognition. *Memory, 10*(1), 1–6.

Polich, J., & Donchin, E. (1988). P300 and the word frequency effect. *Electroencephalography and Clinical Neurophysiology, 70*(1), 33–45.

Polich, J., & Herbst, K. K. (2000). P300 as a clinical assay: Rationale, evaluation, and findings. *International Journal of Psychophysiology, 38*(1), 3–19.

Polich, J., Howard, L., & Starr, A. (1983). P300 latency correlates with digit span. *Psychophysiology, 20*(6), 665–669.

Pope, K. S. (1996a). Science as careful questioning: Are claims of a false memory syndrome epidemic based on empirical evidence? *American Psychologist, 52*(9), 997–1006.

Pope, K. S. (1996b). Memory, abuse and science: Questioning claims about the false memory syndrome epidemic. *American Psychologist, 51*(9), 957–974.

Roediger, H. L., III. (1996). Memory illusions. *Journal of Memory and Language, 35*(2), 76–100.

Roediger, H. L., III, Jacoby, D., & McDermott, K. B. (1996). Misinformation effects in recall: Creating false memories through repeated retrieval. *Journal of Memory and Language, 35*(2), 300–318.

Roediger, H. L., III, & McDermott, K. B. (1995). Creating false memories: Remembering words not presented in lists. *Journal of Experimental Psychology, 21*(4), 803–814.

Rosenfeld, J. P. (1995). Alternative views of Bashore and Rapp's (1993). Alternatives to traditional polygraphy: A critique. *Psychological Bulletin, 117*, 159–166.

Rosenfeld, J. P. (2002). Event-related potentials in the detection of deception, malingering, and false memories. In M. Kleiner (Ed.), *Handbook of polygraph testing* (pp. 265–286). New York: Academic Press.

Rosenfeld, J. P., Angell, A., Johnson, M., & Qian, J.-h. (1991). An ERP-based, control-question lie detector analog: Algorithms for discriminating effects within individuals' average waveforms. *Psychophysiology, 28*, 319–335.

Rosenfeld, J. P., Shue, E., & Singer, E. (2007). Single versus multiple probe blocks of P300-based concealed information tests for self-referring versus incidentally obtained information. *Biological Psychology, 74*, 396–404.

Rosenfeld, J. P., Soskins, M., Bosh, G., & Ryan, A. (2004). Simple, effective countermeasures to P300-based tests of detection of concealed information. *Psychophysiology, 41*(2), 205–219.

Ruchkin, D. S., Sutton, S., Kietzman, M. K., & Silver, K. (1980). Slow wave and P300 in signal detection. *Electroencephalography and Clinical Neurophysiology, 50*, 35–47.

Russell, W. A., & Jenkins, J. J. (1952). *The complete Minnesota norms for responses to 100 words from the Kent-Rosanoff Association Test.* Technical report 11. Minneapolis, MN: University of Minnesota.

Sattler, J. (1998). Reliability, validity, and other considerations related to the interview. In *Clinical and forensic interviewing of children and families: Guidelines for the mental health, education, pediatric and child maltreatment fields* (pp. 206–228). La Mesa, CA: Jerome Sattler.

Schacter, D. L. (1996). *Searching for memory.* New York: Basic Books.

Schacter, D. L., Reiman, E., Curran, T., Yun, L. S., Bandy, D., McDermott, K. B., et al. (1996). Neuroanatomical correlates of veridical and illusory recognition memory: Evidence from positron emission tomography. *Neuron, 17*, 267–274.

Schank, R. C. (1982). *Dynamic memory.* New York: Cambridge University Press.

Schnyer, D. M., Allen, J. J. B., & Forster, K. I. (1997). Event-related brain potential examination of implicit memory processes: Masked and unmasked repetition priming. *Neuropsychology, 11*(2), 243–260.

Semlitsch, H. V., Anderer, P., Schuster, P., & Presslich, O. (1986). A solution for reliable and valid reduction of ocular artifacts, applied to the P300 ERP. *Psychophysiology, 23*(6), 695–703.

Senkfor, A. J., & Van Petten, C. (1998). Who said what? An event-related potential investigation of source and item memory. *Journal of Experimental Psychology: Learning, Memory, & Cognition, 24*(4), 1005–1025.

Spanos, N. P. (1994). Multiple identity enactments and multiple personality disorder: A sociocognitive perspective. *Psychological Bulletin, 116*(1), 143–165.

Spanos, N. (1996). *Multiple identities and false memories* (pp. 105–129). Washington, DC: American Psychological Association.

Spanos, N. P., Burgess, C. A., & Burgess, M. F. (1994). Past-life identities, UFO abductions, and satanic ritual abuse: The social construction of memories. *International Journal of Clinical & Experimental Hypnosis, 42*(4), 433–446.

Spanos, N. P., Burgess, C. A., Burgess, M. F., Samuels, C., & Blois, W. O. (1999). Creating false memories of infancy with hypnotic and non-hypnotic procedures. *Applied Cognitive Psychology*, 13(3), 201–218.

Squires, N. K., Squires, K. C., & Hillyard, S. A. (1975). Two varieties of long latency positive waves evoked by unpredictable auditory stimuli in man. *Electroencephalography and Clinical Neurophysiology, 38*, 387–401.

Sutton, S., & Ruchkin, D. S. (1984). The late positive complex: Advances and new problems. *Annals of the New York Academy of Sciences, 425*, 1–23.

Underwood, B. J. (1965). False recognition produced by implicit verbal responses. *Journal of Experimental Psychology, 70*, 122–129.

Urbach, T. P., Windmann, S. S., Payne, D. G., & Kutas, M. (2005). Mismaking memories: Neural precursors of memory illusions in electrical brain activity. *Psychological Science, 16*, 19–24.

van Hooff, J. C., Brunia, C. H. M., & Allen, J. J. B. (1996). Event-related potentials as indirect measures of recognition memory. *International Journal of Psychophysiology, 21*, 15–31.

Weisbrod, M., Kiefer, M., Winkler, S., Maier, S., Hill, H., Roesch-Ely, D., & Spitzer, M. (1999). Electrophysiological correlates of direct versus indirect semantic priming in normal volunteers. *Cognitive Brain Research, 8*, 289–298.

Woody, C. D. (1967). Characterization of an adaptive filter for the analysis of variable latency neuroelectric signals. *Medical and Biological Engineering, 5*, 539–553.

Zaragoza, M. S., & Mitchel, K. J. (1996). Repeated exposure to suggestions and the creation of false memories. *Psychological Science, 7*(5), 294–300.

SOCIAL NEUROSCIENCE, 2009, 4 (6), 491–509

Psychology Press
Taylor & Francis Group

Thirty-site P300 scalp distribution, amplitude variance across sites, and amplitude in detection of deceptive concealment of multiple guilty items

Ming Ann Lui

Northwestern University, Evanston, IL, USA and National University of Singapore, Singapore

J. Peter Rosenfeld

Northwestern University, Evanston, IL, USA

Andrew H. Ryan, Jr.

Department of Defense Polygraph Institute, Fort Jackson, SC, USA

Previous tests of P300 in deception detection have focused mainly on amplitude analysis. Since countermeasures for such tests have been reported, we looked here at other possible variables as deception indices: P300 scalp distribution and amplitude variance, both across 30 sites. We were also concerned, for the first time, in testing for recognition of more than one guilty item in a mock crime scenario. There were three groups: (1) two-probe group, two of six items were guilty knowledge (GK) items; (2) three-probe group, three of six items were GK items; (3) control group, zero of six items were GK items. In group analyses, in the two-probe group, P300s for lies were significantly greater than P300s for truthful responses. There were significant interactions of condition (Lie vs Truth) by site, suggesting different scalp profiles for deceptive versus truthful responding. Amplitude variance across sites was also greater in Lie than in Truth blocks. These results did not obtain in the three-probe and control groups. In terms of amplitude variances in probe conditions across groups, two-probe group was larger than three-probe and control groups. Regarding individual diagnostics, the variance method yielded a greater-than-chance detection rate of 71% versus 28% false positives. Regarding amplitude at multiple nonfrontal sites, 71% of guilty subjects were detected versus 14% false positives. Grier's (1971) A' indices of various test discrimination efficiencies varied from .76 to .87. Results of the present study suggested further investigation of the variance method as a diagnostic tool for lie detection.

INTRODUCTION

We have previously reported on numerous occasions that the scalp distribution of the P300 event-related potential (ERP) varies depending on whether or not the subject is telling the truth (Miller & Rosenfeld, 2004; Miller, Rosenfeld, Soskins, & Jhee, 2002; Rosenfeld et al., 1998, 1999; Rosenfeld, Rao, Soskins, & Miller, 2002, 2003—the rationale for expecting such effects is also given in these papers). However, these demonstrations all involved group effects, and practical detection of deception requires *intraindividual diagnosis.* This was not accomplished with a scalp distribution index of deception until the report of Rosenfeld (2002), which was promising in that the best distribution index of several that were explored found that 73% of

Correspondence should be addressed to: Peter Rosenfeld, Department of Psychology, Northwestern University, 2029 Sheridan Road, Evanston, IL 60208, USA. E-mail: jp-rosenfeld@northwestern.edu

This research was supported by an award from the Department of Defense Polygraph Institute, DODP198-P-0001.

www.psypress.com/socialneuroscience DOI:10.1080/17470910802047964

the guilty subjects were correctly diagnosed with no false positives. Although a 73% detection rate is not as high as one would like—simple P300 amplitude at Pz as an index of concealed knowledge has been reported to correctly diagnose 85–95% of subjects—there are two reasons why the scalp distribution approach seemed worth following up.

First, Mertens and Allen (2008) and Rosenfeld, Soskins, Bosh, and Ryan (2004) reported that simple countermeasures can dramatically reduce the diagnostic accuracy of the simple P300 amplitude method of detecting deception. We reasoned in Rosenfeld et al. (2004) that since the antecedents for P300 production were relatively well known, it would be easy to train subjects to generate P300s to irrelevant stimuli simply by having them learn covert responses to these irrelevant frequent stimuli. This would turn them into P300-evoking targets. Such a strategy, of course, obscured the difference between P300s to the guilty knowledge "probe" items and to the irrelevant items. On the other hand, the antecedent conditions for altering scalp distribution in a deception detection model are relatively unknown; thus, we reasoned that countermeasures would be difficult to mount against scalp distribution indexes.

Second, in the Rosenfeld (2002) study, the numbers of items out of a total of nine possible on which the subject could have been guilty was not controlled. Some subjects were guilty of one item, which was therefore rarely presented, whereas others were guilty of two or more items. Moreover, different subjects were typically guilty of different items or item subsets, another possible source of error variance. Thus, in diagnosing individual subjects, we were far from dealing with a homogeneous sample, and the sample we studied was small. The present study was in part aimed at testing the scalp distribution method on homogeneous subject groups. Moreover, the present study utilized a mock crime scenario in a concealed information test (CIT) model, rather than an analog screening comparison question test (CQT) as in Rosenfeld (2002); the former allows for better control of subjects' experiences during the run. In the present study, the experimental guilty groups are first run through a truth-telling block, then a deception block. Effects of time passage between the blocks are obtained with a control group, as noted below.

A related but separate issue addressed by the present study concerns the matter known in the polygraph literature as the "multiple issue" question. In the present context, this simply means that there remains a question, never systematically addressed before, as to how well any of the P300-based deception tests perform when the subject is guilty of more than one item in the total tested list, thus increasing the "oddball" probability. In our previous P300-based laboratory analogs utilizing simple Pz amplitude as a recognition index (of denied knowledge in the CIT protocol; e.g., Rosenfeld et al., 2004), the subject was guilty of only one item per test block. Thus the guilty knowledge items had probabilities of .11 to .17 over various studies. Since there are now many such published studies from many labs, in the present study the situation is manipulated so that the subject is guilty of either two (two-probe group) or three (three-probe group) of six tested items (guilty item $p = .33$ and .50 respectively). Our expectation was that since P300 amplitude is determined in part by stimulus probability and meaningfulness (Rosenfeld, Biroschak, Kleschen, & Smith, 2005, and papers therein cited), there would probably be acceptable detection rates, based on simple Pz P300 amplitude with two but not three meaningful items of six items in all. (We note that in terms of numbers of *salient or meaningful items*, one needs to include the target. Thus the numbers above would change to .5 from .33, and .67 from .5 in the two-probe and three-probe groups respectively.)

We also had here a third group (control group) whose members were all innocent with respect to all items. These participants, who did no stealing and therefore no lying, were run through blocks 1 and 2 as above for two reasons: (1) to control for order effects or passage of time between truth-telling (always the first block run) and lying blocks, a fixed order which we felt preferable to counterbalance in a deception situation; (2) to provide an estimate of false positives.

METHODS

Two-probe group

Participants

Northwestern University undergraduate students (14, average age 19.2 years, 4 men) participated in the present study for the fulfillment of an introductory psychology course requirement. They were all right-handed and had normal or corrected vision. All signed an IRB-approved consent form.

Procedures

A mock crime scenario paradigm (e.g., Rosenfeld et al., 1988) was used in this study. Participants were asked, prior to the first running block, to look at five objects in a box, simply to familiarize them with the stimuli to be used in the two blocks to come. We also wanted to equalize experience across groups preceding the first and second blocks. The participant was then asked to imagine that he/she was in a police station, sitting in front of a police officer who suspected theft. Each subject was innocent of theft at this point and was wrongly suspected as an offender. The subjects were then to pretend that the imagined police officer asked him/her to participate in a lie detection test designed to explore whether he/she was guilty or innocent. Each participant was asked to try his/her best to imagine the scenario and play the role he/she was acting.

Next, each participant was seated in a dim room, facing a computer monitor approximately 1 m from his/her eyes. He/she was asked to look at the screen and, when signaled, to respond out loud to each item. A test word, which may have been one of the objects found in the box, would appear. These included: Camera, Recorder, Telephone, Diskman, and Speaker. There was another item he/she would see that was not in the box. It was "Palm", which referred to an electronic organizer Palm Pilot. Then a blank screen would appear briefly, and was followed by the message "Have you stolen this?". Each subject was asked to give a verbal response, repeating the word according to what he/she saw previously on the screen, followed by a "no" which indicated truly that he/she had not stolen the item. When the stimulus was "Palm", subjects were instructed to say "target" rather than "palm". This was to make sure they were paying attention on all trials. There were 12 practice trials for each participant. We will refer to stolen items as probes, other items in box as irrelevants, and the "Palm" items as targets. Note that there were six items in total were presented to all subjects during the test: the five items in the box, and the target item. Two or three (in two-probe and three-probe groups, respectively) items were probes, with three or two (in two-probe and three-probe groups, respectively) items irrelevant, and one item was the target in both groups. Thus the probe probabilities in two-probe and three-probe groups were respectively 2/6 = .33 and 3/6 = .5. In the control group, there were no probes, five irrelevants, and one target.

After the first block, participants were left alone, and each of them was asked to remove (i.e., pretend to steal) two objects from the five objects in a box. As instructed, he/she then hid the objects in a pocket, purse or backpack.

In the subsequent second block, each subject was asked to imagine that he/she was again in a police station, sitting in front of a police officer who suspected that he/she had stolen something. In fact this time he/she had "stolen" two objects from the box. He/she was suspected of the thefts. The police officer again asked him/her to participate in a lie detection test so as to determine whether he/she was guilty or innocent. The task was the same as that in the first block. He/she was alerted that if he/she was guilty, his/her brain waves would probably give him/her away. He/she should try to keep calm; if he/she beat the test, he/she would receive a $20 reward.

Stimulus presentation timing

The stimulus presentation was as follows: The computerized EEG system started at time zero to collect data. First there was a pre-stimulus interval of 104 ms. This served as the baseline EEG recording. In a random fashion, one of the six stimulus words then appeared on the center of the screen for 400 ms as the recording continued simultaneously. (The presentations were actually quasi-random to be sure that the stimuli did not repeat more than once.) The message "Have you stolen this?" then appeared 1544 ms later, just after the EEG recording was finished (total recording epoch = 104 + 400 + 1544 = 2048 ms; in other words, the epoch extended from –104 to 1944 ms, with stimulus onset at 0 ms). The message appeared on the screen for 5.4 s during which time participants gave the verbal response during the first second of the 5.4 s interval (see Figure 1). The long delay was designed to minimize the chance that any speech artifacts were carried over to the next trial. Verbalization rather than button pressing was used since earlier autonomic nervous system (ANS)-based studies reported that greater responses were elicited by verbalized than by non-verbalized reactions (Bradley, MacLaren, & Carle, 1996; Gustavson & Orne, 1965; Rosenfeld et al., 2003). No behavioral responses were recorded by the

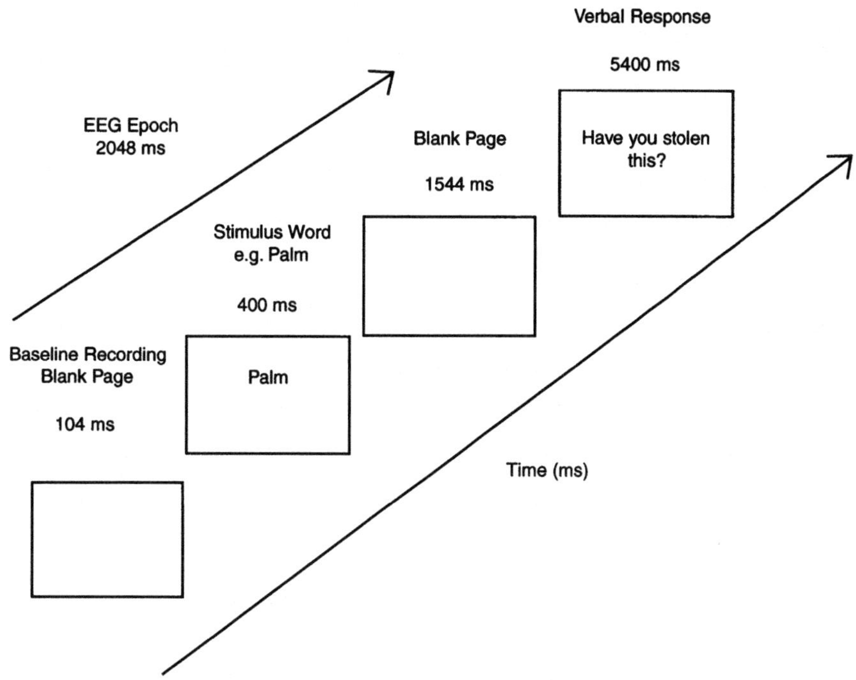

Figure 1. An illustration of an experimental trial. First, there was a pre-stimulus interval of 104 ms as the baseline EEG recording. In a random fashion, one of the six stimulus words then appeared on the center of the screen for 400 ms as the recording continued simultaneously. The message "Have you stolen this?" appeared 1544 ms later, just after the EEG recording was finished (total recording epoch = 104+400+1544 = 2048 ms; i.e., the epoch extended from −104 to 1944 ms, with stimulus onset at 0 ms). The message appeared on the screen for 5.4 s; participants gave the verbal response during the first second of the 5.4 s interval.

computer for the verbal responses made. However, the experimenter made sure that the subjects responded "Target, no" to the "Palm" stimuli in the practice trials. Subjects were asked to redo the practice if they made many errors. And during the real experimental trials, the subjects' responses were monitored by the experimenter. The total inter-stimulus interval (ISI) was around 7.45 s. Each stimulus word was repeated 24 times, so there were 144 trials in each block.

EEG recordings

The EEG data were referentially recorded from 30 tin electrodes in an Electrocap. The reference electrode was put on the nose tip, with the forehead connected to the chassis of the isolated side of the amplifier system (ground). A 0.3 Hz high-pass and a 30 Hz low-pass filter were used (Contact Precision Instruments EEG 8 system). The sampling rate was 125 Hz, and the EEG signal was amplified by a factor of 50,000. EOG was recorded differentially from two electrodes diagonally placed above and below the left eye, so as to monitor both vertical and horizontal eye movement. Trials containing 80 μV or more deflections in the EOG electrodes were automatically rejected (without subject's knowledge). Also, offline visual inspection was done on individual trials to remove trials with especially subtle eye movement artifacts. All electrode resistances were maintained at or below 5 kΩ.

For all group analyses and displays, single sweeps and averages were digitally filtered offline to remove higher frequencies; 3 db point = 4.23 Hz. P300 was measured with the base to peak method: The algorithm searches within a window from 400 to 1000 ms post-stimulus for the maximally positive segment average of 104 ms. The pre-stimulus 104 ms average is also obtained and subtracted from the maximum positivity to define the measure. The midpoint of the maximum positivity segment defines P300 latency.

Three-probe group

Participants

Northwestern University undergraduate students (17, average age 18.8 years, 7 men) participated in the present study for the fulfillment of an introductory psychology course requirement. They were all right-handed and had normal or corrected vision. All signed IRB-approved consent forms.

Procedures

The experimental procedures in this three-probe group were basically the same as those used in the two-probe group described above, except that the participants were asked to remove three (as opposed to two in the two-probe group) of the five objects in the box between the two blocks. All other methods (including EEG data collection and statistical analysis methods) were the same as in the two-probe group.

Control group

Participants

Northwestern University undergraduate students (14, average age 19.6, 7 men) participated in the present study for the fulfillment of an introductory psychology course requirement. They were all right-handed and had normal or corrected vision. All signed IRB-approved consent forms.

Procedures

The experimental procedures in this control group were basically the same as those used in the two-probe and three-probe paradigms described above, except that the participants were asked to simply take a break between the first and second blocks; i.e., no stealing was done. All other methods (including EEG data collection and statistical analysis methods) were the same as in the two-probe group and three-probe group.

Analyses

The type of data we were interested in analyzing is illustrated by Figure 2 from the two-probe group, which shows P300 amplitude as a function of site for the original 30-site set, in the four

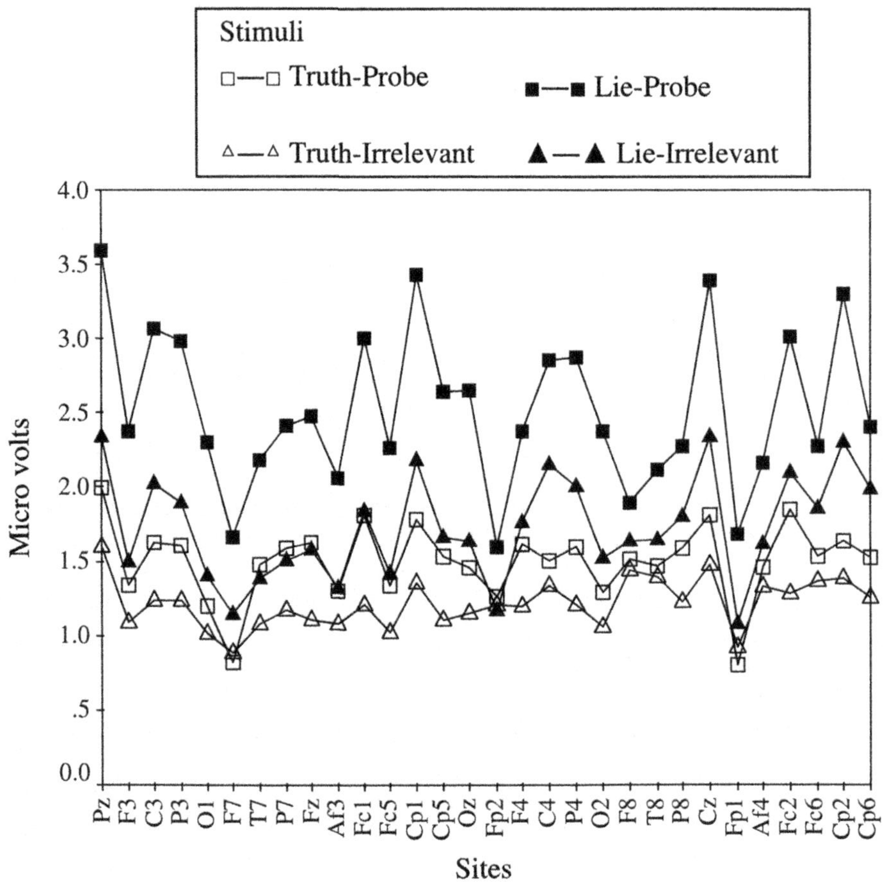

Figure 2. Plot of calculated P300 amplitude (μV) as a function of four conditions and 30 scalp sites in two-probe group.

conditions of the experiment: (1) Truth-Probe, (2) Truth-Irrelevant, (3) Lie-Probe, and (4) Lie-Irrelevant. The first two conditions refer to probe and irrelevant responses during the first block when all subjects told the truth; the latter two conditions refer to the same response types during the second block when subjects in the two-probe group and three-probe group lied. Of course, in block 1, which preceded the mock thefts, all stimuli are irrelevant; "Truth-Probe" therefore denotes those items to be stolen just prior to the second block. For the two-probe group and the three-probe group, all responses to the two or three probe items, respectively, are combined and all responses to the three or two irrelevant items, respectively, are combined, for group and individual tests to be described below. The sets of two or three probe items averaged in the two-probe or three-probe group were the items selected by each individual. For the control group, stimuli are all irrelevant in both blocks. For calculation of false positives, the same two randomly selected (by experimenter) items were designated as probes and three as irrelevants for both blocks. Different items were selected across individuals. Thus the control group controlled for effects reported below in the two-probe group. There was no need to control for three-probe group results since no group effects were seen, as reported below. We do not present target data as these are of no present interest and showed no effects anyway.

Group analyses

The first appropriate analyses here were omnibus analyses of variance (ANOVAs); four conditions (as above and as in Figure 2) by 30 sites, followed up with key contrasts of interest. Main effects of amplitude and site were expected in the two-probe group, since probe amplitudes in the Lie block should have exceeded those in the Truth block inasmuch as only in the Lie blocks for the two-probe group were the probes relatively rare and meaningful. We also anticipated a Condition by Site interaction in comparing scalp distributions on the expectation that lie distributions should look different than truth distributions (see Miller et al., 2002; Miller & Rosenfeld, 2004; Rosenfeld et al., 1998, 1999, 2002, 2003). We expected site effects in the three-probe group, but could not honestly anticipate other effects. A main condition effect might be expected on the basis of the fact that probes were especially meaningful, but at $p = .5$, they were not rare, so that only one of the two critical antecedent conditions for P300 elicitation was present for certain. We expected no parallel effects for the control group, where all stimuli but targets were equivalent and irrelevant.

We note that the key contrasts of interests in the two-probe group and three-probe group concern comparison of (1) Truth-Probe vs Lie-Probe, (2) Lie-Probe vs Lie-Irrelevant. If the subject is lying only to the probes representing stolen items in the second block, the Lie-Probe P300 should be larger than both the Truth-Probe and Lie-Irrelevant responses.

In this study, in group and individual analyses, we compared unscaled scalp distributions, since we are making no inferences about localization of neurogenerators, but only wished to develop indices for detecting deception. Also, Urbach & Kutas (2002) cast some doubt on the validity of scaling, which remains controversial. This is further considered in the discussion.

We also note that for group analyses, all 30 sites were used. We used the principal component analysis method described below to reduce the dimensionality for individual analyses, where the bootstrap methods for scalp distributions, also described below, would be very time-consuming with all 30 sites in each subject.

Intraindividual analyses

It is noted that when group effects were not seen, no individual effects were analyzed except in the control group, in which the aim in part was to estimate a false positive rate. Three intraindividual tests were performed.

Bootstrapped amplitude difference (SIZE)

To determine whether or not the P300 evoked by one stimulus is greater than that evoked by another within an individual, the bootstrap method (Wasserman & Bockenholt, 1989) is usually used on the Pz site where P300 is typically largest. This will be illustrated with an example of a probe response being compared with an irrelevant response. The question answered by the bootstrap method is: "Is the probability more than 90 in 100 that the true difference between the average probe P300 and the average irrelevant P300 is greater than zero?" For each subject, however, one has available only one average probe P300 and one average irrelevant P300. Answering the statistical question requires distributions of average P300 waves, and these actual distributions are not available. One thus bootstraps the distributions, in the bootstrap variation used here, as follows. A computer program goes through the probe set (all single sweeps) and draws at random, with replacement, a set of n1 waveforms. It averages these and calculates P300 amplitude from this single average using the maximum segment selection method as described below. Then a set of n2 waveforms is drawn randomly with replacement from the irrelevant set, from which an average P300 amplitude is calculated. n1 is the actual number of accepted probe sweeps for that subject, and n2 is the actual number of accepted irrelevant sweeps for that subject. The calculated irrelevant mean P300 is subtracted from the comparable probe value, and one thus obtains a difference value to place in a distribution that will contain 100 values after 100 iterations of the process just described. Multiple iterations will yield differing (variable) means and mean differences due to the sampling-with-replacement process.

In order to state with 90% confidence that probe-evoked and irrelevant-evoked ERPs are indeed different, one requires that the value of zero difference or less (a negative difference) not be greater than –1.29 standard deviations from the mean of the difference distribution for 90% confidence. In other words, the lower boundary of the 90% confidence interval for the difference would be greater than zero. It is noted that sampling different numbers of probes and irrelevants could result in differing errors of measurement; however, studies have shown a false positive rate of zero utilizing this method (Ellwanger, Rosenfeld, Sweet, & Bhat, 1996) and others have taken a similar approach (Farwell & Donchin, 1991) with success. This method has the advantage of utilizing all the data, as would an independent groups *t*-test with unequal numbers of subjects. It is further noted that a one-tailed 1.29 criterion yields a $p < .1$ confidence level because the hypothesis that the probe-evoked P300 is greater than the irrelevant-evoked P300 is rejected either if the two are not found significantly different or if the irrelevant P300 is found larger (*t*-tests on single sweeps are too insensitive to use to compare mean probe and irrelevant P300s within individuals; see Rosenfeld, Angell, Johnson, & Qian, 1991).

Individual analysis

Scalp distribution analysis: Spatial principal component analysis followed by novel bootstrap tests

For intra-individual analysis, we also created "virtual sites" with principal component analysis (PCA) as a data reduction method. PCA was applied to deal with the large amount of data collected from the 30 actual channels. PCA on sites in space is a method to identify clusters of electrodes that are highly intercorrelated. It linearly combines the highly correlated electrodes to form a virtual component or factor. The component or factor forms the "virtual site" data to be used in later analyses. This has the advantage of capturing most of the relevant variance and therefore preserving the information from the electrodes, and at the same time reducing the redundancy among them (Spencer, Dien, & Donchin, 2001). Additionally, the linear combination method of PCA can produce orthogonal factors.

An individual PCA was done on each subject's data as each subject might have had a different scalp distribution for reasons related to individual differences. The data set was obtained from an average of ERPs at each time point within a given window of all trials in the same stimulus type. The data input into the covariance-based PCA consisted of 30 electrode sites by a number of observations (number of time points × four condition types as described above, as in Figure 2). The time points included in the two-probe paradigm, the three-probe paradigm and the control group data were within the time window in which the P300 peak was captured in all subjects within each group. As each individual has P300 at a different latency, the time window selected for different groups also differed slightly. The time windows for the two-probe, three-probe and control groups were 480–872 ms, 480–904 ms and 400–936 ms respectively. For individual subject data, Kaiser-Meyer-Olkin measure of sampling adequacy ranged from .743 to .908, which supports that the windows captured enough data for each individual's PCA model.

Spatial factors were extracted for the Varimax rotation until more than 80% of the variance was accounted for. Varimax rotation was chosen because it produces factors with high loadings on a small number of variables, and low loading on other variables (Kaiser, 1960). Also, factors remained uncorrelated after Varimax rotation, which prevents the problem of multicollinearity among factors. Standardized factor loadings were the correlations between variables (original sites) and its corresponding factor (virtual site). A site variable was selected only if its factor loading exceeded 0.6 and if such a high loading for this site variable was restricted to one factor. The factor scores were formed by summing the retained site values multiplied by respective factor score coefficients (Spencer, Dien, & Donchin, 2001). Each individual had a different set of factor scores, and different numbers of virtual sites accounting for 85% or more of the variance. Putting results of the three groups together, two virtual sites were adequate in 14 subjects, three in 20 subjects and for 11 others, four virtual sites were required.

In our study, we intended to find out whether lying and truth-telling subjects had different scalp distribution profiles. We examined whether the curves of average amplitude as f(virtual sites) in the lie-telling condition differed from those in the truth-telling conditions, including probe in the Truth condition and in the Lie condition. This part of the analysis is based on the view that a significant interaction effect in an analysis of variance between two conditions implies a difference in the profiles of the two conditions. We performed Condition × Site ANOVAs to obtain the interaction effects (F of interaction terms) between the lying and truth-telling conditions as described next.

A novel bootstrapping procedure was developed for this analysis. Assume there were x and y single sweeps in the lie- and truth-telling conditions respectively in one subject. In the first part of the algorithms, x and y single sweeps were drawn randomly (with replacement) from the actual pool of data in each of the lie- and truth-telling blocks separately and respectively, and an average was calculated for each randomly selected group. The random selection process was carried out 100 times to get 100 sets of averages in each group. An ANOVA was done on the two groups of randomly selected single sweeps to get an F-value for the interaction between the two groups. The whole random selection and ANOVA process was carried out 10 times to yield 10 F-values of the interaction term. These 10 F-values were averaged to get a grand-averaged F-value for comparison with the following, second part results.

In the second part of the algorithm, the single sweeps from both conditions were shuffled. The same random selection was then done as in the first part, except that the pool of single sweeps to draw from now contained data from both conditions, as a result of shuffling. An F-value was now computed for these two randomly selected groups, and this procedure was carried out 50 times to yield a distribution. The grand-averaged F-value in part 1 was located in the distribution of part 2. If the F-value in part 1 was in the top 10% of the distribution in part 2, we concluded that the scalp distribution profiles between the lie and truth-telling data were dissimilar and the subject was classified as guilty.

Individual analysis

Variance

The aim here was to compute P300 amplitude variance across sites within each condition and then to compare variances between conditions. For example, the graph of the average P300 amplitude over 30 sites in four conditions in Figure 2 seems to show that the variance of the curves of the two conditions in the Lie block was larger (note the ranges) than that of the two conditions in the Truth block. The variance comparison described here aimed at statistically validating this apparent difference. Post-hoc, between-block statistical tests on variance were done. A t-test was done within each subject to compare the difference in variance of mean amplitudes over 30 sites between conditions. Due to the correlation between paired sample data in within-subject design which may cause an underestimation of t-value, an adjustment was made (Fergusson, 1971). The formula is as follows:

$$t = [(s_1^2 - s_1^2) \times (N-2)^{0.5}]/[2 \times s_1 \times s_2 \times (1 - r_{12}^2)^{0.5}]$$

where s_1 and s_2 are the standard deviations of the pair of sample data in the two conditions, r_{12} is the correlation between the pair of data compared, and N is the sample size.

Results for some comparisons of the three-probe group and the control group are also given in terms of the signal detection theoretical parameter A', based on Grier (1971). This is a function of the distance between a Receiver Operating Characteristic (ROC) curve and the main diagonal of an ROC plot of hits and false alarms. It makes no assumptions about the shape or variances of the distributions of the key variables (such as *P–I* P300 amplitude differences).

$$A' = .5 + \{[(y - x) \times (1 + y - x)]/4 \times y \times (1 - x)]\}$$

where y = hit rate; x = false alarm rate.

RESULTS

Figure 3 shows the grand averaged Truth-Probe and Lie-Probe ERPs at Fz, Cz, and Pz for the two-probe and three-probe groups. Lie-Probe averages suggest larger P300s than Truth-Probes in both groups, but the difference appears larger in the two-probe group. Figure 4 shows the same kind of data for irrelevant responses. In the two-probe but not the three-probe group there appear to be block differences. In the three-probe group there are apparent differences in the waveforms after P300, but not during the time window in which one sees P300. (Other ERP components, based on a spatio-temporal PCA as in Spencer et al. (2001), are reported in a companion paper.) Figure 5 shows similar grand averages in the two groups in which probe and irrelevant responses from the second (lie) blocks are superimposed. As expected, there are clear differences restricted to the P300 region in the two-probe group, with much smaller effects for the three-probe group.

Figures 2, 6, and 7 show the calculated P300 amplitudes as a function of condition and site in the two-probe, three-probe, and control groups, respectively. It appears that in the two-probe

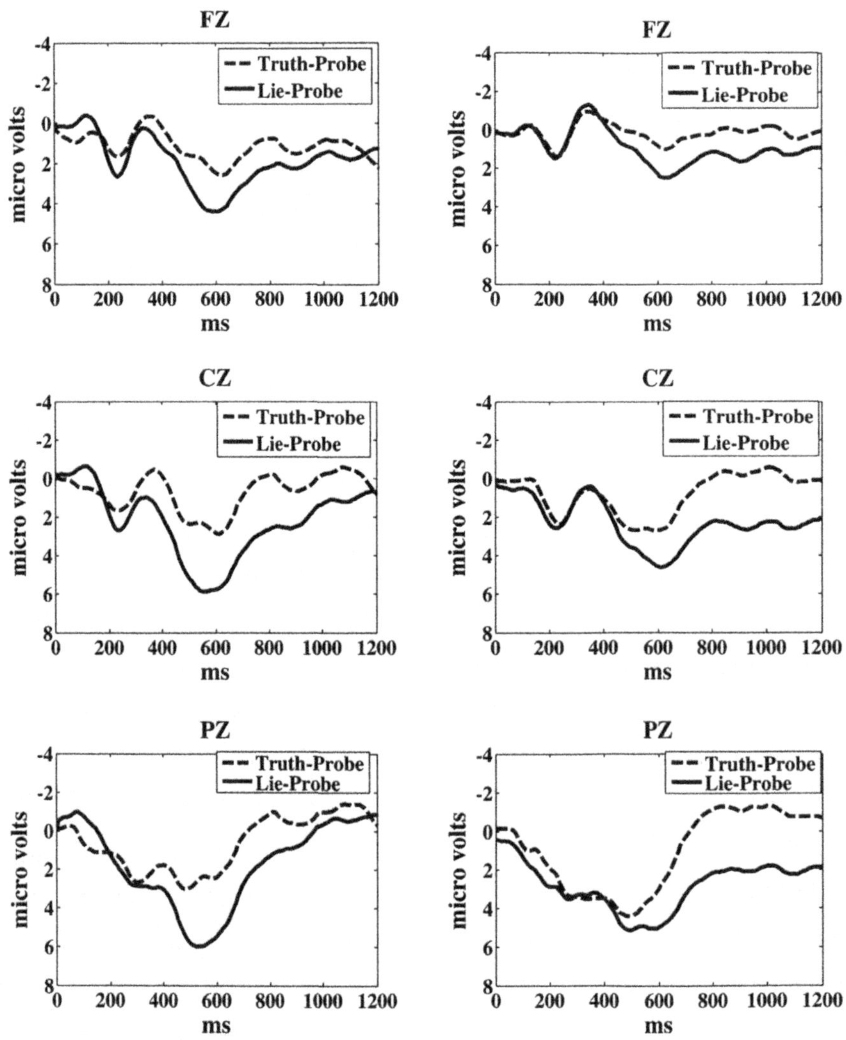

Figure 3. Grand average Truth-Probe and Lie-Probe: ERPs at classic midline sites in two-probe group (left column) and three-probe group (right column), respectively (stimulus onset at 0 ms).

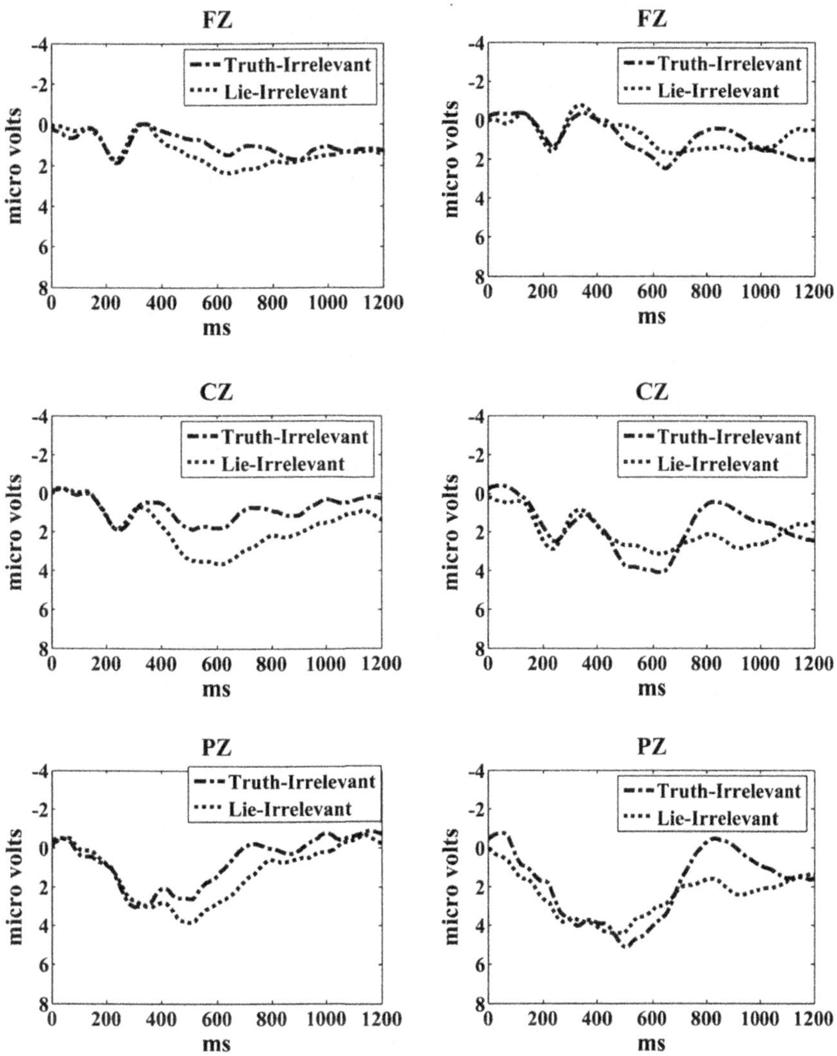

Figure 4. Grand average Truth-Irrelevant and Lie-Irrelevant: ERPs at classic midline sites in two-probe group (left column) and three-probe group (right column), respectively (stimulus onset at 0 ms).

group, the Lie-Probe P300s at many sites are larger than the P300s in the Truth-Probe and Lie-Irrelevant curves, as expected. (These would be the two most appropriate comparisons to examine in individuals in the present paradigm for diagnostic purposes.) The two-probe group also appears to show more variable scalp distributions in the Lie response curves than the Irrelevant curves, suggesting that the Lie distributions and amplitude variances may differ from the others. Figure 6 (three-probe group) does not show clear effects corresponding to those seen in Figure 2 (two-probe group). In Figure 7 (control group), only two conditions are shown corresponding to the first and second blocks. The amplitudes plotted as an f(site) are based on two randomly selected items per block (two pairs for two blocks per subject) since there were no probes vs irrelevants as no items were treated in any special manner for the control group. All items were irrelevant. (The reason for the selection of two rather than three items per block was that this group was meant to control for effects seen in the two-probe group in which two items were stolen by each subject.) The two scalp distributions shown in Figure 7 do not appear to differ appreciably, nor do the mean amplitudes.

To analyze Figure 2 (two-probe group), a 4 (Condition) $\times$ 30 (Site) repeated measure ANOVA was done on the P300 baseline-to-peak amplitudes. The four conditions were probe in the Truth block, irrelevant in the Truth block, probe in the Lie block, and irrelevant in the Lie

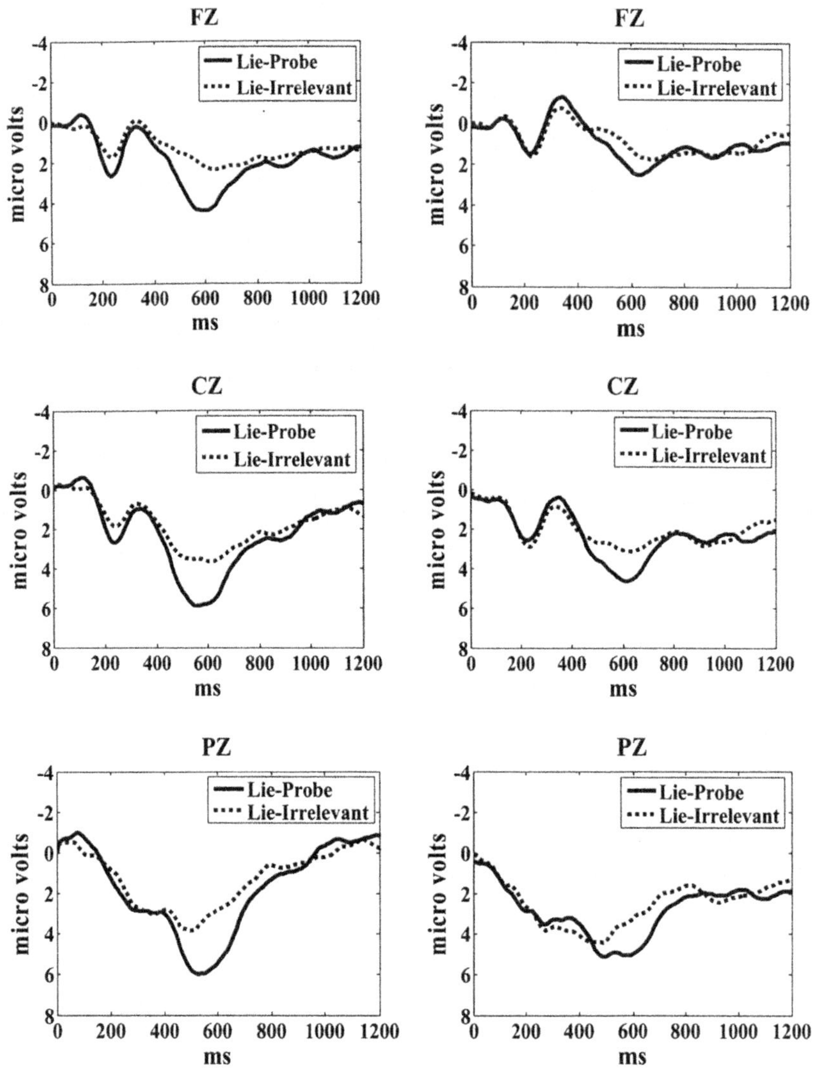

Figure 5. Grand average Lie-Irrelevant and Lie-Probe: ERPs at classic midline sites in two-probe group (left column) and three-probe group (right column), respectively (stimulus onset at 0 ms).

block. We first used Mauchly's test of sphericity to check whether the ANOVA assumption of sphericity was violated. If so, Greenhouse-Geisser correction was applied. There were significant main effects of Condition, $F(3, 39) = 4.482$, $p = .008$, and Site, $F(29, 377) = 3.963$, $p = .008$. There was also a significant interaction between Condition and Site, $F(87, 1131) = 2.234$, $p = .046$, showing that there was a different pattern of the P300 baseline-to-peak amplitude over different sites in different conditions. A follow-up 2×30 (Condition $\times$ Site) ANOVA with only probe in Truth block and probe in Lie block included in the factor of condition was done. As expected, probe in Lie block had a significantly larger P300 amplitude than probe in Truth block $F(1, 13) = 4.722$, $p = .049$, and there was a near significant interaction between the two conditions $F(29, 377) = 2.461$, $p = .06$. There was also a main effect of site, $F(29, 377) = 4.934$, $p = .002$. Another follow-up 2×30 ANOVA was done in which probe and irrelevant in the Lie blocks were the two conditions. There was again an expected main effect of Condition: $F(1, 13) = 5.526$, $p < .04$, a main effect of Site: $F(29, 377) = 5.582$, $p < .001$, but the interaction trend did not reach significance ($p = .182$). Although latency was not of critical or pre-specified interest in this study, a 4×30 Condition $\times$ Site ANOVA was done on the P300 latency. The main effect of Site was significant, $F(29, 406) = 13.468$, $p < .001$. There wasn't a significant effect of Condition, $F(3, 42) = 0.124$, $p = .946$, nor of interaction, $F(87, 1218) = 1.027$, $p = .418$.

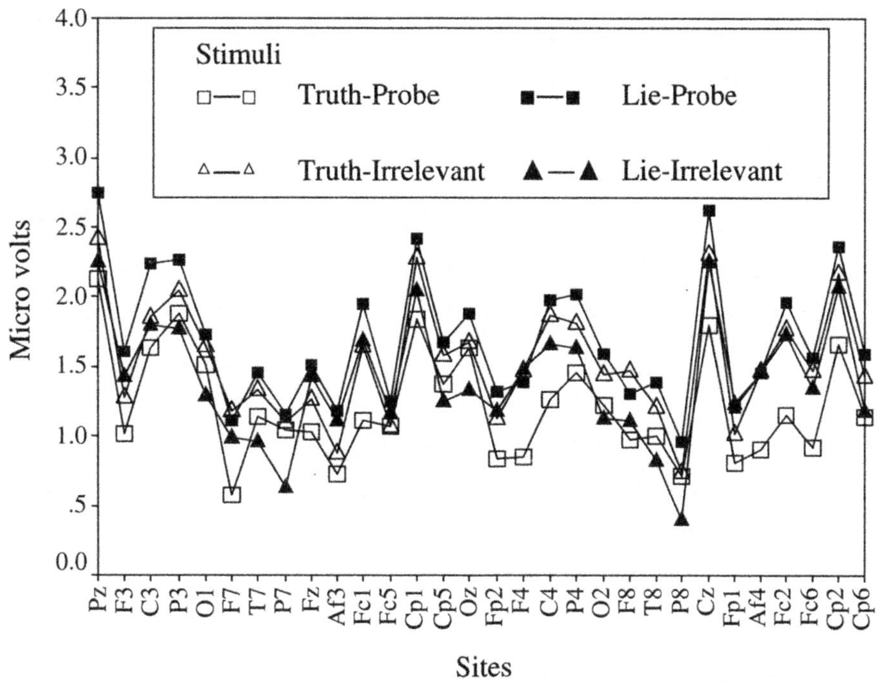

Figure 6. Plot of calculated P300 amplitude (μV) as a function of four conditions and 30 scalp sites in three-probe group.

Group analysis of difference in mean variance of P300 baseline-to-peak amplitudes over 30 sites was done. The mean variance of P300 baseline-to-peak amplitudes for probe in the Lie block = .283 μV^2 was significantly larger than that for the probe in the Truth block = .068 μV^2, $t(29) = 6.866$, $p < .001$, and significantly larger for the probe in the Lie block than for the

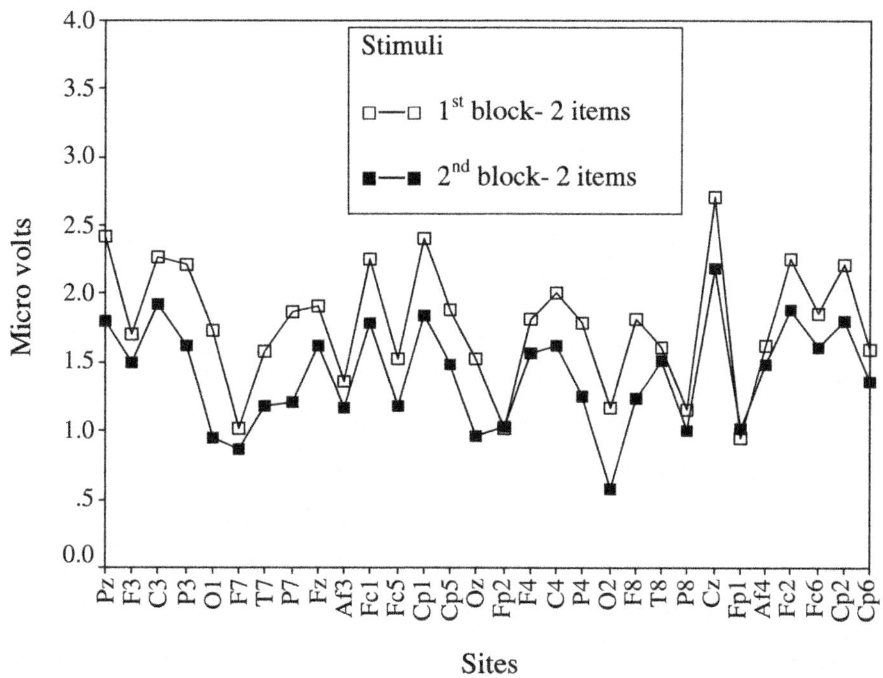

Figure 7. Plot of calculated P300 amplitude (μV) as a function of two conditions (representing time passage) and 30 scalp sites in control group with no probes. The conditions refer to a first block, then a second block, and responses to the same two randomly chosen stimuli are used in both blocks.

irrelevant in the Lie block = .125 μV^2, $t(29) = 4.873$, $p < .002$.

Regarding three-probe group (Figure 6), a 4 × 30 Condition by Site repeated measures ANOVA was done on the P300 baseline-to-peak amplitudes. The main effect of Site was significant, $F(29, 464) = 5.649$, $p = .01$. The main effect of Condition was not significant, $F(3, 48) = 0.783$, $p = .509$. The interaction between Condition and Site was not significant either, $F(87, 1392) = 1.296$, $p = .256$. Since the main effects of condition and interaction were not significant, we did not follow up with 2 × 30 ANOVAs, as we did do in the two-probe group.

Group analysis of difference in mean variance of P300 baseline-to-peak amplitudes over 30 sites was done. The mean variance of P300 baseline-to-peak amplitudes for probe in the Lie block (.224 μV^2) was significantly larger than that for the probe in the Truth block (.156 μV^2), $t(29) = 2.414$, $p < .05$, while there was no significant difference between the probe in the Lie block and the irrelevant in the Lie block (.197 μV^2), $t(29) = .847$, $p > .05$.

Regarding the control group, for all the group analyses, two irrelevant items were randomly selected (so as to allow comparison with Experiment 1) in each block for each individual subject. A 2 × 30 Condition by Site repeated measure ANOVA was done on the P300 baseline-to-peak amplitudes as shown in Figure 7. The two conditions were two randomly selected items in the first block and in the second block. As no stealing was done, all items were irrelevant stimuli. The main effect of Site was significant, $F(29, 377) = 3.481$, $p = .021$. The main effect of Condition, as expected, was not significant, $F(1, 13) = 0.525$, $p = .482$, nor, as expected, was the interaction effect between Condition and Site, $F(29, 377) = 0.904$, $p = .464$.

Group analysis of difference in mean variance of P300 baseline-to-peak amplitudes over 30 sites was done. The mean variance of P300 baseline-to-peak amplitudes for two randomly selected items in the second block (.141 μV^2) was significantly smaller than for the two randomly selected items in the first block (.204 μV^2), $t(29) = 2.134$, $p < .05$; note that this direction of difference is opposite to what was seen with the two-probe group regarding block order. This suggests that without the mock crime, there is some reduction of variance across scalp sites with passage of time, and that the significant difference seen in the two-probe and three-probe groups was partially cancelled by the time passage, yet was seen despite this effect. There was no significant difference between the mean variance of the two randomly selected items and that of the three remaining items in the second block (.189 μV^2), $t = 1.903$, $p > .05$.

Cross-group comparisons

Besides the effects already reported within the groups, we expected Lie-Probe P300s in the two-probe group to have larger P300 amplitudes and possibly different distributions in comparison with comparable variables in the other groups in Lie blocks. A cross-group analysis of difference in mean P300 baseline-to-peak amplitudes was done in two key conditions. For the control group, again, two randomly selected items in the first block and two randomly selected items in the second block were used; for the two-probe and three-probe groups, probe in the Truth block and probe in the Lie block were used as two conditions in the first analysis. In a second analysis, probe in the Lie block and irrelevant in the Lie block constitute the two conditions. Amplitudes were compared among the control group, two-probe group and three-probe group. The 2 × 30 × 3 (condition × site × group) mixed design ANOVA has "Condition" and "Site" as the within-subject variables and "Group" as the between-subject factor.

In the first analysis, the main effect of Condition was not significant, but a trend was suggested; $F(1, 42) = 2.671$, $p = .110$. The main effect of Site was significant, $F(29, 1218) = 10.985$, $p < .001$. The main effect of Group was not significant, $p > .4$. The interaction between Condition and Group was close to significant, $F(2, 42) = 2.863$, $p = .068$, which, again, suggested that the effect of Condition on the P300 amplitudes tended to differ in different groups. The other two-way interactions, Group × Sites ($p = .582$) and Condition × Sites ($p = .131$), were not significant. The three-way interaction, Condition × Sites × Group, was close to significant, $F(58, 1218) = 1.697$, $p = .079$.

In the second, cross-group, 2 × 30 × 2 analysis in which Lie-Probe and Lie-Irrelevant were the two conditions used in two-probe group, there was a main effect of Condition: $F(1, 42) = 4.219$, $p < .05$, and there was also a main effect of Site: $F(29, 1218) = 13.278$, $p < .01$, and there were trends for the interactions of Condition × Group: $F(2, 42) = 2.278$, $p = .115$, and the triple interaction,

$F(58, 1218) = 1.503$, $p = .159$, but the other interactions and main effect of group were $p > .2$.

To focus on the group contrast we expected to be clearest, the three-probe group was removed from the group comparison, and two randomly selected items in the second block of the control group and the probe in the Lie and Truth blocks in the two-probe group were compared with a first follow-up analysis: a $30 \times 2 \times 2$ (Site by Group by Condition) mixed design ANOVA in terms of the mean P300 baseline-to-peak amplitudes. There was a main effect of Site, $F(29, 754) = 7.268$, $p < .001$. The effect of Group was now marginally significant, $F(1, 26) = 3.761$, $p = .063$. The effect of Condition was not significant, $p > .3$. The interaction Site × Group was not significant, $p > .7$. The interaction of Condition × Group was marginally significant at $F(1, 26) = 4.038$, $p < .06$. The Condition × Site interaction was not significant, $p > .4$. The triple interaction (Condition × Group × Site) was significant, $F(29, 754) = 2.496$, $p < .04$.

In the second follow-up analysis without the three-probe group, the probe and irrelevant in the Lie block in the two-probe group were compared with two remaining randomly selected items in the control group. The main effect of the Condition trend did not reach significance, $F(1, 26) = 2.811$, $p = .106$. There was a main effect of Site, $F(29, 754) = 8.388$, $p < .001$, but not Group, $p > .18$. There was a significant interaction of Condition × Group, $F(1, 26) = 4.345$, $p < .05$ and trend for the triple interaction, $p = .15$. But the other interactions of sites × group and condition × sites were not significant at $p > .4$.

A final set of cross-group analyses of difference among variances over 30 sites was also done. The variance for the probes in the Lie block (.283 μV^2) of the two-probe group was significantly higher than that of the probe in the Lie block (.224 μV^2) for the three-probe group, $t(29) = 3.065$, $p < .01$. Nevertheless, this variance for the probe in the three-probe group was significantly larger than that for two randomly selected items in the control group (1.41 μV^2), $t(29) = 3.565$, $p < .01$. It follows that the variance for the probe in the two-probe group is larger than that in the control group, and this is confirmed by analysis: $t(29) = 5.891$, $p < .001$.

Individual analyses: Pz/Cz amplitudes and variance, virtual scalp distributions

These analyses were done only in the two-probe and control groups, since there were no group effects of condition nor condition × group in the three-probe group, and we thus decided that there was no point in looking for individual effects. We did analyses in the control group so as to estimate false positive effects. Tables 1 and 2 summarize the results for two-probe and control groups respectively. We can state here that many of the results were disappointing, but this is tempered by

TABLE 1
Two-probe paradigm within-subject analysis results

Subject	*1*	*2*	*Either 1 or 2*	*3*	*4*	*5*	*Either 4 or 5*
1							
2	*	*	*	*	*		*
3					*		*
4	*		*	*	*	*	*
5							
6				*		*	*
7	*		*				
8		*	*			*	*
9	*	*	*	*		*	*
10	*	*	*		*	*	*
11					*	*	*
12							
13				*		*	*
14					*		*
Detection	5/14 (35.7%)	4/14 (28.6%)	6/14 (42.9%)	5/14 (35.7%)	6/14 (42.9%)	7/14 (50%)	10/14 (71.4%)

Notes: *A significant result. Column headings: (1) bootstrapping on baseline to peak amplitude (Truth-Probe vs Lie-Probe); (2) bootstrapping on baseline to peak amplitude (Lie-Probe vs Lie-Irrelevant); (3) Bootstrapping on interaction effects; (4) variance difference between probe in Truth block and probe in Lie block; (5) variance difference between probe and irrelevant in Lie block.

TABLE 2
Control group within-subject analysis results

Subject	*1)*	*2)*	*Either 1) or 2)*	*3)*	*4)*	*5)*	*Either 4) or 5)*
1							
2				*			
3	*			*			
4							
5	*			*			
6							
7				*	*	*	*
8				*			
9							
10				*			
11					*	*	*
12						*	*
13		*				*	*
14							
Detection	2/14 (14.3%)	1/14 (7.1%)	3/14 (21.4%)	6/14 (42.9%)	2/14 (14.3%)	4/14 (28.6%)	4/14 (28.6%)

Notes: *a significant result (false positive); (1) bootstrapping on baseline to peak amplitude (two-item 1st block vs two-item 2nd block); (2) bootstrapping on baseline to peak amplitude (two-item 2nd block vs three-item 2nd block); (3) bootstrapping on interaction effects; (4) variance difference between two items in 1st block and two items in 2nd block; (5) variance difference between two items in 2nd block and the remaining three items in Lie block.

the encouraging results seen in variance data, and in amplitude difference data on all real sites (versus virtual sites); see below.

For the two-probe group (see Table 1), a bootstrapping procedure was done on the analysis of baseline-to-peak amplitude difference at Pz and Cz sites. Four out of 14 subjects (28.6%) showed significantly larger amplitude at both Pz and Cz for the probe in the Lie block than for the irrelevant in the Lie block. Five of 14 subjects (35.7%) showed significant differences in comparisons of Truth-Probe vs Lie-Probe responses. Six out of 14 subjects (42.9%) showed a significant difference in either one or the other of these two comparisons. There were no other significant differences in the other subjects. These disappointing individual diagnostic results with subjects guilty of two probes (probe $p = .33$) do not compare well with results from subjects in previous studies guilty of only one probe item (probe $p = .17$; e.g., Rosenfeld et al., 2004), and are therefore followed up in the next section by doing individual analyses at all 30 sites.

To get at scalp distribution differences, a bootstrapping procedure was applied to each subject in two-probe group virtual site data (as described in 'Methods' above) extracted with a PCA performed within each subject. In 5 out of 14 (35.7%) subjects, there were significant interaction effects in ANOVAs of the comparison between probe in the Truth block and probe in the Lie block. There were no significant effects in the remaining subjects. Note in Table 1 that only 3 out of 14 subjects have differences in both amplitude and distribution tests.

The following comparisons are on the variance of mean P300 amplitudes over 30 sites in two-probe group (see Table 1). For both Truth-Probe vs Lie-Probe and Lie-Probe vs. Lie-Irrelevant comparisons, detection is poor; however, 10/14 (71.4%) subjects have either larger variance for probe in the Lie block than for probe in the Truth block, or larger variance for probe in the Lie block than for irrelevant in the Lie block. This compares favorably with the false positive rate of 4/14 (28.6%), estimated on the same either-or criterion on variance, in the control group (Table 2), as described below. A Fisher Exact test comparing these two independent proportions yielded $p < .03$, and the Grier (1971) A' parameter evaluating discriminative efficiency was .8.

Although the lack of significant group effects precluded within-subject analysis for three-probe group, we do perform these intra-individual tests in control group so as to estimate false positive rate. (See Table 2 for the following results.) The bootstrapping procedure was done on the analysis of baseline-to-peak amplitude difference. Two out of 14 subjects (14.3%) had significantly larger amplitude at both Pz and Cz for two randomly selected items in the second block than that in the first block. One out of 14 subjects (7.1%) got significantly larger amplitude at both Pz and Cz for two randomly selected items in the second block than for the three remaining items in the second block. Three out of 14 subjects (21.4%) showed a significant difference in either one of these two comparisons. This false positive rate is about half that seen in two-probe group (but A' will not be computed with insensitive tests).

A bootstrapping procedure was done on the virtual site data that were produced by a PCA. As Table 2 (control group) shows, in 6 out of 14 (42.9%) subjects, there were significant interaction effects in ANOVAs done for the comparison between items in the first and second blocks.

For each control group (Table 2) subject, trials of two items from each block were randomly selected. The mean variance of P300 amplitudes over 30 sites for items chosen in the first block was compared to that of the second block. Four out of 14 (28.6%) subjects got either significantly larger variance for randomly selected items in the first block than for randomly selected items in the second block, or larger variance for randomly selected items in the second block than for nonselected items in the second block.

Individual analyses: Amplitude effects at actual sites

Since the results of amplitude analysis with just Pz and Cz were disappointing, and since variance data across the 30 actual sites were suggestive, we decided to examine (as a final intra-individual analysis) Truth-Probe versus Lie-Probe amplitudes (Contrast 1) and Lie-Probe vs Lie-Irrelevant P300s (Contrast 2) at all 17 nonfrontal members of the original set of actually recorded 30 sites in all subjects of two-probe group and control group. (Frontal sites were omitted in this analysis since P300s tend to be most prominent more posteriorly.) Summary results are given in Table 3. A significant bootstrapped amplitude difference result in the expected direction (Lie-Probe > Truth-Probe; Lie-Probe > Lie-Irrelevant, both in guilty Ss) in any set of nonfrontal sites was taken as a correct detection. It is seen that in the two critical comparisons, the proportions of detected subjects in the two-probe group were two to five times greater than comparable proportions from the control group. Fisher exact tests of two-probe group versus control group proportions yielded $p = .13$ for Truth-Probe versus Lie-Probe ($A' = .76$) and $p < .03$ for Lie-Probe versus Lie-Irrelevant, yielding $A' = .87$. Combining proportions on both contrasts yielded $p < .008$ ($A' = .82$). It is noted that while a hit was declared if only one non-frontal site tested positively, in many cases there were multiple sites showing the effect within each two-probe group subject; for Contrast 1, 11 two-probe group subjects showed effects at >5 (of 17) sites. For Contrast 2, 11 two-probe group Ss showed effects in >5 sites. There was not, however, a highly consistent set of sites showing effects with either contrast across subjects, as can be seen in Table 4, where results at all 30 sites are shown in the leftmost two columns for the two-probe group.

TABLE 3
Numbers of subjects showing predicted directional P300 bootstrapped amplitude differences at 1 or more sites (not including frontal sites) for Truth-Probe (Block 1) versus Lie-Probe (Block 2) conditions, and Lie-Probe versus Lie-Irrelevant conditions (both Block 2)

Group	*Truth-Probe vs Lie-Probe*	*Lie-Probe vs Lie-Irrelevant*
Two-probe	9/14 (64%)	10/14 (71%)
Control	4/14 (29%)	2/14 (14%)

Looking at all 30 sites, for Contrast 1, the mean number of sites per subject across all two-probe group Ss showing significant expected guilty effects was 9.7, versus 2.8 in the control group, $t(26) = 1.926$, $p < .07$). For Contrast 2, the mean number of sites per subject across all two-probe group Ss showing significant expected guilty effects was 9.9, vs 1.4 in the control group, $t(26) = 2.611$, $p < .02$). If the mean site numbers for the two contrasts are added, then the summed sites means for the two-probe group and control group are 19.6 and 4.2, respectively, $t(26) = 2.75$, $p < .02$.

To summarize results of individual subject analysis on sites, it appears that deception involving more than one probe item results in larger probe P300s at various sites (Table 4), and not, as has been seen previously with one probe, reliably at the classic Cz and Pz sites.

DISCUSSION

The results of group analyses performed separately within each group produced expected results based on the well-known probability effect in the P300 literature. That is, the two-probe group with two probes among six items showed

TABLE 4
Numbers of subjects diagnosed guilty based on both Truth-Probe versus Lie-Probe amplitudes (Contrast 1) and Lie-Probe vs Lie-Irrelevant P300s (Contrast 2) at all sites

	Two-probe group		*Control group*	
Site	*Contrast 1*	*Contrast 2*	*Contrast 1*	*Contrast 2*
fp1	5	4	3	1
fp2	4	4	3	0
af3	3	4	1	1
af4	4	3	1	0
fz	4	5	1	1
f3	5	4	2	1
f4	5	2	2	0
f7	5	4	1	1
f8	1	2	2	1
t7	5	7	1	0
t8	3	3	1	0
fc1	5	6	1	1
fc2	4	4	3	1
fc5	4	6	1	1
fc6	4	2	1	0
cz	5	4	1	1
c3	7	6	1	1
c4	5	4	1	1
cp1	6	7	1	2
cp2	5	6	1	1
cp5	4	7	1	1
cp6	3	4	1	1
pz	6	6	1	1
p3	6	6	1	1
p4	6	4	1	1
p7	3	6	1	0
p8	4	4	1	0
oz	6	5	1	0
o1	5	5	1	0
o2	4	5	1	0

significant effects of Condition, with Lie-Probe amplitudes > Truth-Probe amplitudes, and Lie-Probe amplitudes > Lie-Irrelevant amplitudes. So-called Truth-Probe trials were equivalent to Irrelevant trials in either block prior to the mock crime. These effects were not seen in the three-probe group (nor in the control group where there were no probes or oddballs). This is as expected based on the known inverse effect of oddball probability on P300 amplitude (e.g., Rosenfeld et al., 2005, and papers therein cited). In the present two-probe group, the probe (oddball) probability was .33, vs .5 in three-probe group; indeed, it is probably not appropriate to call an item with $p \geq .5$ an oddball. Regarding the multiple issue problem noted in the introduction, these data suggest that to successfully use multiple probe items in a P300-based CIT, the total number of items must be increased so as to maintain an acceptably low oddball probability. However, we did show for the first time here that individual diagnostic indices (discussed below) can be developed that show fairly good discrimination of truth-tellers and deceivers when there are two probe items with combined $p = .33$. Certainly, A' values should exceed .9 for field use, so we are not there yet. (If one takes the target into account—see above—then the picture here improves in that these methods work with three salient of six total items; two probes plus one target.) In previous studies with only one probe item, its probability was typically .17, one of six total items.

In contrast, previous P300-based CITs utilizing single probes (Allen et al., 1992; Farwell & Donchin, 1991; Rosenfeld et al., 1988, 1991, 2004) typically reported guilty detection rates of 80–95%, utilizing only the Cz or Pz sites in bootstrap analysis. In the present study, even when restricting analyses to the two-probe group, the simple bootstrapped amplitudes using only Pz or Cz could not detect even 50% of the guilty subjects. Indeed, even when using the complex indices discussed below, the best we could detect in the two-probe group did not exceed 71% of the guilty subjects. This number compared favorably to false positive estimates based on the same indices for the control group, but it is still not as large as most agencies would want. Perhaps it could be increased by lowering probe probabilities via manipulation of total numbers of items as noted above. This hope rests on the suggested interpretation that the poor results seen here with analyses on single sites were due to the higher probe probabilities utilized.

We were also interested here in extending our earlier studies (cited in the introduction) of scalp distribution as a deception index. Utilizing condition by site interaction terms to index different distributions between conditions, as previously, the best group result we found was a strong trend ($p < .06$) in the two-probe group for the P300 distribution in the Lie-Probe condition to differ from the Truth-Probe condition. There were a few other marginally significant interactions ($.05 < p < .1$) involving site in the cross-group comparisons, but nothing robust. This was somewhat surprising since several previous studies (cited in the introduction) reported clearly significant group effects regarding condition by site interactions. These earlier studies, however, were not mock crime studies, and their oddball probabilities were half that used here. This probability factor would be one to suspect as a possible source of the difference between previous and present results in view of the possibility of effects of event probability on P300 scalp distribution. Also, we have shown that information type also affects P300 amplitude in CITs (Rosenfeld et al., 2004; Rosenfeld, Biroschak, & Furedy, 2006), and the mock crime studies as utilized here contain different types of probe information than the self-referring and numerical probe information used in previous scalp distribution studies. It is reasonable to suspect that the protocol effect might also account for the conflicting results.

Another measure of amplitude distribution that we used here for the first time was variance across differing scalp sites. This index proved to be quite sensitive in group analyses. In both the two-probe and three-probe groups, the Lie-Probe variance was greater than the Truth-Probe variance, and in the two-probe group, additionally, the Lie-Probe variance was greater than the Lie-Irrelevant variance. Moreover, in direct cross-group comparison, the Lie-Probe variance of the two-probe group was greater than that of the three-probe group, which was greater than that for the comparable comparison in the control group. This variance index was thus sensitive to the oddball probability difference between the two-probe and three-probe groups, and was the only index to show any significant difference within the three-probe group indicating that Lie-Probe P300s differed from Truth-Probe P300s.

Careful study of Figures 2 and 6 suggests why the variance index was more successful than the Condition × Site interaction index. It is apparent that most of these plots are similar and show peaks at the same sites, but the filled squares in Figure 6 indicating the Lie-Probe data often show higher maxima and lower minima than those seen in other plots. This greater range in the Lie-Probe curve is the probable source of the variance effect when Lie-Probe variance is compared with other variances. The fact that the curves are geometrically similar suggests that the distributions are mostly similar, so there would be few Condition × Site interactions. Indeed, one would expect that if these distributions were scaled in some way (McCarthy, & Wood, 1985) there would be no condition by site interactions found, and not yet reported here, that absence of interactions in scaled data is exactly what we observed. Thus, the expectation that deception and truth-telling would engage differing neurogenerator populations as indexed by differing scaled scalp distributions (Rosenfeld, 2002; Rosenfeld et al., 1998) is not supported by the present results, and it becomes necessary to speculate about the meaning of the variance effects. The individual diagnostic amplitude data may provide a hint: By looking within individuals, we found here that subjects in deceptive relative to truthful conditions showed increased probe amplitudes at various sites in lie trials. This is inferred from the results of bootstrapped difference tests as in Tables 3 and 4. (A positively diagnosed subject at a given site is showing a probe > irrelevant result, suggesting an increase in probe amplitude.) However, in contrast to the case when the probe probability is low as in previous studies, in the present study the increased probe amplitude is not consistently seen at the sites (Cz, Pz, etc.) where P300 is typically largest, but is seen at site sets that differ across subjects, and that do not always contain the classic P300 sites. It is as if in the present protocol, deceptive probe trials will tend to increase P300 amplitude at least *at some, but not all, locations*, if not the classic sites. If such is the case, then within a given subject one cannot expect an increased Lie-Probe P300 at one (consistently classic) location, but one can expect elevations of amplitude in some and not all locations. This effect could be captured by amplitude variance across the scalp in most subjects, even if the sites with increased amplitudes are not the same across subjects. Discovering the physiological mechanism of this effect of deception on amplitude variance across the scalp is a question for future research.

REFERENCES

Allen, J., Iacono, W.G., & Danielson, K.D. (1992). The identification of concealed memories using the event-related potential and implicit behavioural measures: A methodology for prediction in the face of individual differences. *Psychophysiology, 29*, 504–522.

Bradley, M. T., MacLaren, V. V., & Carle, S. B. (1996). Deception and non-deception in guilty knowledge and guilty actions polygraph tests. *Journal of Applied Psychology, 81*, 153–160.

Ellwanger, J. W., Rosenfeld, J. P., Sweet, J., & Bhat, M. (1996). Detecting simulated amnesia for autobiographical and recently learned information using the P300 event-related potential. *International Journal of Psychophysiology, 23*, 9–23.

Farwell, L.A., & Donchin, E. (1991). The truth will out: Interrogative polygraphy ("lie detection") with event-related potentials. *Psychophysiology, 28*, 531–547.

Fergusson, G. A. (1971). *Statistical analysis in psychology and education* (3rd ed., p. 167). New York: McGraw-Hill.

Grier, J. B. (1971). Non-parametric indexes for sensitivity and bias: Computing formulas. *Psychological Bulletin, 75*, 424–429.

Gustavson, L. A., & Orne, M. T. (1965). The effects of verbal responses on the laboratory detection of deception. *Psychophysiology, 2*(1), 10–13.

Kaiser, H. F. (1960). The application of electronic computers to factor analysis. *Educational and Psychological Measurement, 20*, 141–151.

Lui, M., Rosenfeld, J. P. (in press). Detection of deception about multiple, concealed, mock crime items, based on a spatial-temporal analysis of ERP amplitude and scalp distribution. *Psychophysiology.*

McCarthy, G., & Wood, C. (1985). Scalp distributions of event-related potentials: An ambiguity associated with analysis of variance models. *Electroencephalography and Clinical Neurophysiology, 62*, 203–208.

Mertens, R., & Allen, J. J. (2008). The role of psychophysiology in forensic assessment: Deception detection, ERPs, and virtual reality mock crime scenarios. *Psychophysiology, 45*, 286–298.

Miller, A. R., & Rosenfeld, J. P. (2004). Response-specific scalp distributions in deception detection and ERP correlates of psychopathic personality traits. *Journal of Psychophysiology, 18*, 13–26.

Miller, A. R., Rosenfeld, J. P., Soskins, M., & Jhee, M. (2002). P300 amplitude and topography in an autobiographical oddball paradigm involving simulated amnesia. *Journal of Psychophysiology, 16*, 1–11.

Rosenfeld, J. P. (2002). Scaled P300 scalp profiles in detection of deception. *Technical Report to DOD for Project DoDPI98-P-0001*. Jackson, SC: DoDPI Press.

Rosenfeld, J. P., Angell, A., Johnson, M., & Qian, J. (1991). An ERP-based, control-question lie detector analog: Algorithms for discriminating effects within individuals' average waveforms. *Psychophysiology, 38*, 319–335.

Rosenfeld, J. P., Biroschak, J. R., & Furedy, J. J. (2006). P300-based detection of concealed autobiographical versus incidentally acquired information in target and non-target paradigms. *International Journal of Psychophysiology, 60*(3), 251–259.

Rosenfeld, J. P., Biroschak, J. R., Kleschen, M. J., & Smith, K. M. (2005). Subjective and objective probability effects on P300 revisited. *Psychophysiology, 42*, 356–359.

Rosenfeld, J. P., Cantwell, B., Nasman, V. T., Wojdac, V., Ivanov, S., & Mazzeri, L. (1988). A modified, event-related potential-based guilty knowledge test. *International Journal of Neuroscience, 42*, 157–161.

Rosenfeld, J. P., Ellwanger, J. W., Nolan, K., Wu, S., Bermann, R. G., & Sweet, J. J. (1999). P300 scalp amplitude distribution as an index of deception in a simulated cognitive deficit model. *International Journal of Psychophysiology, 33*, 3–19.

Rosenfeld, J. P., Rao, A., Soskins, M., and Miller, A. R. (2002). P300 scalp distribution as an index of deception: Control for task demand. *Journal of Credibility Assessment and Witness Psychology, 3*(1), 1–22. Retrieved March 19, 2008, from http://truth.boisestate.edu/jcaawp/200201/200201.htm

Rosenfeld, J. P., Rao, A., Soskins, M., & Miller, A. (2003). Scaled P300 scalp distribution correlates of verbal deception in an autobiographical oddball paradigm: Control for task demand. *Journal of Psychophysiology, 17*, 14–22.

Rosenfeld, J. P., Reinhart, A. M., Bhatt, M., Ellwanger, J., Gora, K., Sekera, M., et al. (1998). P300 correlates of simulated amnesia on a matching-to-sample task: Topographic analyses of deception vs. truth-telling responses. *International Journal of Psychophysiology, 28*, 233–248.

Rosenfeld, J. P., Soskins, M., Bosh, G., & Ryan, A. (2004). Simple effective countermeasures to P300-based tests of detection of concealed information. *Psychophysiology, 41*, 205–219.

Spencer, K. M., Dien, J., & Donchin, E. (2001). Spatiotemporal analysis of the late ERP responses to deviant stimuli. *Psychophysiology, 38*, 343–358.

Urbach, T. P., & Kutas, M. (2002). The intractability of scaling scalp distributions to infer neuroelectric sources. *Psychophysiology, 39*, 791–808.

Wasserman, S., & Bockenholt, U. (1989). Bootstrapping: Applications to psychophysiology. *Psychophysiology, 26*, 208–221.

SOCIAL NEUROSCIENCE, 2009, 4 (6), 510–517

Psychology Press
Taylor & Francis Group

Can simultaneously acquired electrodermal activity improve accuracy of fMRI detection of deception?

F. Andrew Kozel
University of Texas Southwestern Medical Center, Dallas, TX, USA, and Medical University of South Carolina, Charleston, SC, USA

Kevin A. Johnson
Medical University of South Carolina, Charleston, SC, USA

Steven J. Laken
Cephos Corporation, Pepperell, MA, USA

Emily L. Grenesko
Medical University of South Carolina, Charleston, SC, USA

Joshua A. Smith
Medical University of South Carolina, Charleston, SC, USA

John Walker
Medical University of South Carolina, Charleston, SC, USA

Mark S. George
Medical University of South Carolina, Charleston, SC, USA

Observation of changes in autonomic arousal was one of the first methodologies used to detect deception. Electrodermal activity (EDA) is a peripheral measure of autonomic arousal and one of the primary channels used in polygraph exams. In an attempt to develop a more central measure to identify lies, the use of functional magnetic resonance imaging (fMRI) to detect deception is being investigated. We wondered if adding EDA to our fMRI analysis would improve our diagnostic ability. For our approach, however, adding EDA did not improve the accuracy in a laboratory-based deception task. In testing for brain regions that replicated as correlates of EDA, we did find significant associations in right orbitofrontal and bilateral anterior cingulate regions. Further work is required to test whether EDA improves accuracy in other testing formats or with higher levels of jeopardy.

Correspondence should be addressed to: F. Andrew Kozel, M.D., UT Southwestern Medical Center, Department of Psychiatry, 5323 Harry Hines Blvd, Dallas, TX 75390-9119, USA. E-mail: Andrew.Kozel@utsouthwestern.edu

Funding provided by Defense Agency for Credibility Assessment (formerly Department of Defense Polygraph Institute) (W74V8H-04-1-0010) and Cephos Corporation. F. A. Kozel is supported by a K23 from the NIMH (5K23MH070897). We wish to thank Dave Ramsey, Scott Geer, and John Dornish of the South Carolina Research Authority for computer support; Paul Morgan and Chris Rorden for help with imaging parameters; and Minnie Dobbins and Sally Donegan for administrative support. The views expressed in this article are solely those of the authors and do not necessarily reflect the official policy or position of the Department of Defense, the US Government, the National Institute of Mental Health, or the National Institutes of Health.

www.psypress.com/socialneuroscience DOI:10.1080/17470910801907168

INTRODUCTION

Some of the earliest forms of lie detection involved measuring peripheral autonomic arousal in order to detect if someone was being deceitful (Furedy, 1986). The most commonly employed tool for lie detection today, the polygraph, still relies on changes in measures of arousal to attempt to distinguish deceptive versus honest answers (Geddes, 2002). The polygraph and other measures of peripheral arousal (voice stress analysis, thermal imaging, etc.), however, are not measuring deception itself, but a potential consequence of deception. Imaging brain changes associated with deceptive versus truthful answers offers the ability to directly detect deception in the brain where the lie was created. A number of groups have investigated the neurocircuitry associated with various forms of deception (Johnson, George, & Kozel, 2007). Paradigms have included lying about location of money (Kozel, Padgett, & George, 2004a; Kozel et al., 2004b), which object was taken (Kozel et al., 2005), which playing card one is holding (Langleben et al., 2002, 2005; Phan et al., 2005), autobiographical information (Abe, Suzuki, Mori, Itoh, & Fujii, 2007; Abe et al., 2006; Ganis, Kosslyn, Stose, Thompson, & Yurgelun-Todd, 2003; Nunez, Casey, Egner, Hare, & Hirsch, 2005; Spence et al., 2001), memory impairment (Lee et al., 2002, 2005), and staged firing of a weapon (Mohamed et al., 2006). Most of these studies have simply looked at group differences, but two studies have achieved detection of deception at the individual level (Kozel et al., 2005; Langleben et al., 2005).

In the Kozel et al. (2005) study (see paper for complete details), 30 healthy adults (Model Building Group) were studied in order to develop a blood oxygen level-dependent (BOLD) fMRI method of detecting deception that was 93% accurate. An additional independent sample of 31 subjects (Model Testing Group) was recruited and tested to determine how well our automated analysis would work. Using an *a priori* defined methodology, we were able to achieve 90% accuracy (Kozel et al., 2005).

For the Model Testing Group (MTG), simultaneous physiologic measures were acquired during the fMRI testing of deception. The acquired EDA (a.k.a. galvanic skin response), pulse, and respiratory values are indices of autonomic arousal that are similar to the measures for a polygraph test. Because the question format in our study was fundamentally different from the format in which a polygraph test is administered, we were unable to use the physiologic data in a manner similar to a polygraph. Because EDA is one of the primary channels used by polygraphers to detect deception, however, we were interested as to whether adding EDA as a covariate in our testing model would improve or worsen our diagnostic ability. Adding the EDA as a regressor potentially would account for variance or confounds in the data that reduced the accuracy of the fMRI method. Thus, if this were the case, adding the EDA as a covariate would improve the diagnostic ability of the methodology. Conversely, if the EDA is highly correlated with the brain BOLD signal changes associated with our method of detecting deception, adding the EDA as a covariate of non-interest may reduce the power of our method to detect deception. This would be similar to the effect of regressing out head motion in a paradigm in which head motion is highly correlated with the event of interest. Also, we wished to determine whether we could replicate our prior results of EDA being significantly correlated with areas of the brain involved with deception including the right orbitofrontal/inferior frontal, right middle frontal, and right cingulate gyrus (Kozel et al., 2004b).

METHODS

fMRI detection of deception paradigm

The subjects for this analysis consisted of the second cohort of participants (MTG) who were recruited specifically to test our previously developed methodology of detecting deception. Details of the methods and results for the two groups with regard to functional MRI detection of deception are presented elsewhere (Kozel et al., 2005). Briefly, participants were 18–50 years old and healthy as defined by a Structured Clinical Interview for DSM-IV Axis I Disorders (SCID-I) (First, Spitzer, Williams, & Gibbon, 1995), a pre-MRI screening form, a medical history, and a physical exam. Additionally, an Annett Handedness Scale (Annett 1970), a urine drug test, and a urine pregnancy test (if a female of childbearing potential) were performed. Participants were recruited from the Medical University of South Carolina community at large. Informed written consent approved by the Medical University of South Carolina's Office of Research Integrity was obtained prior to any study procedures. Subjects were excluded for taking medications/drugs, being pregnant, or having a medical or psychiatric illness. Subjects were shown the questions that they would be asked during the scanning day. Participants who qualified for the study were scheduled for MRI scanning at a subsequent time.

On the scanning day, participants were taken to a room near the scanner and instructed to "steal" a watch or a ring located in a drawer. The subjects placed the "stolen" object in a locker along with their other belongings. The questions were practiced prior to entering the MRI. Using an Integrated Functional Imaging System (IFIS) (MRI Devices Corp., Gainesville, FL), questions were visually displayed in the scanner and button press responses were recorded. During the fMRI deception testing, yes/no questions were asked that involved the "ring" (regarding whether they took the ring) and the "watch" (regarding whether they took the watch); also asked were "neutral" questions (general questions with clear "yes" and "no" answers) and "control" questions (about doing minor wrongful acts). Participants were instructed to answer the control and neutral questions truthfully, and the ring and watch questions as if they had stolen neither object. Thus, subjects were lying regarding one of the objects and telling the truth regarding the other object. To provide an incentive for subjects to attempt countermeasures, participants were informed that they would receive an additional $50 if investigator FAK could not tell when they were lying while being observed answering questions in the scanner. All subjects received the additional $50. After scanning, subjects completed a questionnaire about the study.

Scanning was performed on a 3T MRI scanner (Intera, Philips Medical System, The Netherlands) using an eight-channel SENSE head coil. The Deception task involved acquiring 515 echo-planar imaging (EPI) transverse images (TR 1867 ms, TE 30 ms, flip angle 90°, FOV 208 mm, matrix 64×64, SENSE factor 2, 36 slices, 3 mm with 0 mm gap, giving a voxel size of $3.25 \times 3.25 \times 3.00$ mm^3) that covered the entire brain and were positioned with reference to the anterior commissure–posterior commissure (AC–PC) line using a sagittal scout image.

For the Deception task, there were 20 yes/no questions for each category (80 unique questions in total). The order of the type of questions was consistent across subjects, but the order of the actual questions asked was varied. The 80 questions were presented once for the practice outside the scanner and then in two separate sets administered consecutively for the imaging run.

Acquiring EDA data

The EDA was acquired using an MRI compatible system developed by MRA Incorporated (Washington, PA). Two electrodes (EL508, BIOPAC Systems, Goleta, CA) were placed on the palm, roughly one over the abductor pollicis brevis muscle and one over the abductor digiti minimi brevis muscle. Unshielded RT Pinch leads (Lead108, BIOPAC Systems, Goleta, CA) were connected to the electrodes, and the signal was transferred out of the scanning environment using a custom LC (inductor capacitor) filter system tuned to suppress the interference from the scanner (MRA Inc., Washington, PA). The signal was then calibrated using a custom Wheatstone bridge built by JW. All signals were collected using National Instruments hardware (SCB-68, DAQCard-6062E, and 68-Pin cable, National Instruments, Austin, TX) and recorded using National Instruments software (LabVIEW Version 8, National Instruments, Austin, TX). The recording range for the filtered and calibrated EDA signal was -2 to 2 V. The EDA value was zeroed at the beginning of each scan and acquired every 0.01 s. The LabView program began acquiring data at the beginning of the first scan (first TTL pulse from scanner) and terminated after the last scan was completed.

Data analysis

In order to use the EDA values as covariates in our model as well as to look for correlates of EDA and BOLD signal, an EDA value was required that corresponded to each whole brain volume. Using Matlab R2006a (Mathworks, Natick, MA) and a custom script developed by SJL, the 96,000 EDA values were normalized [normalized$(t1) = (t1 - \text{mean}(t1))/\max\{\text{abs}[t1 - \text{mean}(t1)]\}$] for each subject and then were averaged for every 1.867 s corresponding to each TR. Thus, there were 515 normalized mean EDA values for each subject.

The first data analysis was performed in the same manner as in the prior study. Briefly summarized, the analysis of the functional MRI data was performed using Statistical Parametric Mapping software (SPM 2, Wellcome Department of Cognitive Neurology, London, UK) run on 2006a

(Mathworks, Natick, MA). Preprocessing of the functional MRI data included images being reoriented to match the SPM2 EPI template, realigned and unwarped, slice time corrected, spatially normalized to the SPM EPI template, resampled with a voxel size of $3 \times 3 \times 3$ mm (Ashburner & Friston, 1999), and then spatially smoothed using a Gaussian kernel with 8 mm full width at half maximum (Friston, 2004). A general linear model within SPM2 was specified and estimated to create individual *t*-maps for the contrasts of Lie –Neutral and Truth–Neutral. The six motion recorded parameters generated during the "Realign" process, and the non-protocol events were modeled as well. A high pass filter (cutoff frequency = 128 s) was used to remove possible effects of low-frequency changes. The number of significant voxels ($p <$ 0.05, uncorrected) for each contrast was determined in the three regions of interest (roughly the right orbitofrontal/inferior frontal, the right middle frontal, and the right anterior cingulate) previously defined. If the number of significant voxels was greater for the Lie–Neutral contrast versus the True–Neutral contrast, then it was considered a correct call. Conversely, if the True–Neutral contrast had a greater number of significant voxels than the Lie–Neutral contrast, then it was considered an incorrect call. If the two contrasts had equal number of significant voxels, then an indeterminate call was made.

To directly compare whether the addition of EDA improved accuracy, the second analysis used the same procedure except that the EDA was added as a regressor in the general linear model. The number of correct calls was compared between the two methods. Also, whether the new model increased the robustness of the difference between the two contrasts was examined.

Despite the questioning format being significantly different from a typical polygraph exam, the EDA data were tested alone to determine if they were able to successfully differentiate deception. A custom Matlab script developed by JAS was used to normalize (same normalization as above) the 96,000 EDA data points and extract them into separate vectors based on question type (lie, true, control, neutral). For each question, 600 data points corresponding to the 6 s question period were extracted and placed into the appropriate vector. The script was used to further calculate a mean EDA for each question period and an overall mean for each question type.

Data generated using this script were exported to the SPSS software package (SPSS for Windows, Version 12, Chicago, SPSS, Inc.) in order to perform statistical analyses. First, SPSS was used to calculate a group average for the lie, truth, control, and neutral conditions. The group average EDA for the lying condition was then used to perform a paired *t*-test to compare the means and to determine an appropriate threshold EDA value to use for prediction of deception. Because a gradual decrease in EDA values was noted during nearly every scanning session, a Wilcoxon signed ranks test was used to compare questions of similar time periods (e.g. lie 1–true 1, lie 4–condition 4). This was done in order to correct for the decrease in EDA values, which made analysis of overall-session average EDA for each condition difficult. For each subject, we compared lie–true, lie–neutral, and true–neutral using a Wilcoxon signed ranks test. A *Z* value and *p* value for each Wilcoxon signed ranks test was determined, and these values were averaged for the entire group.

SPSS was utilized to perform binary logistic regression modeling on the average EDA data. Our model compared full-session average EDA for the lie and truth conditions and used a dichotomous variable with a value of either 0 or 1 to distinguish between the two. SPSS generated an estimated coefficient, a standard error, Wald statistic, and *p*-value for the binary logistic regression.

The final analysis involved testing for associations between EDA and BOLD fMRI signal that was independent of the deception testing. EDA was modeled as a single regressor with the motion parameters included as regressors of non-interest. The purpose of the motion parameters was to reduce the likelihood that any significant area observed was not due to motion. The individual contrast maps of EDA were incorporated into a random effects group map and a total effect of interest was determined using an *F*-test. The statistical threshold was defined as false discovery rate (FDR) of $p < 0.005$ with a minimum cluster size (k) of 20 voxels to help correct for multiple comparisons. Using MRIcro (Version 1.39), the anatomical brain regions and Brodmann areas (BA) corresponding to the cluster's highest statistical threshold voxel were determined. In addition, the entire anatomic area of each cluster was determined using MRIcro.

RESULTS

Subjects

The MTG had 31 subjects scanned. Of these subjects, five had their EDA acquired on different equipment. Since the comparability of the two devices has not been established, only the 26 subjects with EDA acquired on the MRA Inc. equipment are reported. This subset of the MTG had a mean age of 33.0 years (SD 9.8, range 18–50 years). Fifteen of the 26 participants were women, 21 were right-handed, 3 left-handed, and 2 were ambidextrous. The majority of the group was employed full-time (19 participants), with one being part-time employed and six being students. The sample was racially diverse with 14 participants Caucasian, 10 African-American, and 2 Asian-American.

Effect of EDA as regressor in model

Using the standard method, 24 of 26 (92%) were called correctly. With the EDA added as a regressor, 24 of 26 (92%) were called correctly. One person who was previously called correct became incorrect, and one person who was called incorrect became correct. Using the EDA as a regressor, 9 participants had their calls strengthened (i.e. greater difference in the number of active voxels in the Line – Neutral versus the True – Neutral), 12 had their calls weakened (inverse of strengthened), and 5 had no difference.

EDA alone to identify deception

Mean EDA as a predictor

No significant difference was noted in group mean EDA for lie and truth questions (lie $= -0.00273$ V, SD $= 0.01810$; true $= -0.00146$ V, SD $= 0.02043$; paired t-test, $t = -0.197$, $p = 0.845$). Although there was no significant difference between mean EDA during lying and truth-telling events, we attempted to use the group average EDA during lying as a threshold for predicting deception. Using a threshold of 0 V, we were able to correctly predict which item 14 of the 26 subjects took (53.85% accuracy). We were unable to use any non-subjective value to accurately predict deception any better than chance.

Wilcoxon signed ranks test

As a gradual decrease in EDA values was observed throughout nearly every session, we compared questions similar in time frame to correct for this drift. Comparison of lie–true, lie–neutral, and lie–control questions at comparable times yielded no significant results. The group average Z value for the lie–true contrast was –1.114, with a corresponding average p value of 0.412. The lie–neutral contrast was also not statistically significant, with a group average Z value of –1.372 (average p value $= 0.302$). Finally, the overall group average Z value for the lie–control contrast was calculated to be -1.113, with a corresponding p value of 0.382.

Binary logistic regression

A binary logistic regression was also used to analyze mean EDA for the lie and true conditions. The results of the model are $B = -3.553$, standard error 14.676, Wald statistic 0.059, with p value 0.0809. The calculations generated using the binary logistic regression model also suggest that no significant difference between the deceptive and truth-telling events occurred in EDA values. Our model was not able to predict deception better than chance (accuracy $= 50.0\%$).

Correlates of EDA with BOLD fMRI signal

The group map of EDA revealed significant associations in several regions that were found in our previous analysis (See Table 1, Figure 1). Specifically, the replicated regions with significant association include the right orbitofrontal, the right and left anterior cingulate, the right inferior temporal, the right caudate, and the right middle temporal. In addition for this analysis, strong activations were seen in the right and left insula. Interestingly, none of these brain areas associated with EDA overlapped with the fMRI detection of deception methodology regions.

DISCUSSION

The use of EDA in our model of fMRI lie detection did not significantly alter the accuracy of the methodology. The number of correct calls did not change with the addition of EDA as a

TABLE 1

Group analysis of EDA correlated with BOLD fMRI signal: areas of significant activation (false discovery rate <0.005, k ≥20)

Cluster	*k*	*Voxel p(FDR-cor)*	*Voxel F*	*Voxel equivZ*	*MNI coordinates of voxel with largest F-value*			*Anatomic location of voxel with largest F-value*	*Brodmann's area of voxel with largest F-value*	*Complete anatomic area of cluster*
1	172	0	75.26	5.72	−39	−12	12	Left insula	48	Left insula
		0.002	30.48	4.27	−51	−21	27	Left postcentral gyrus	48	Left superior temporal
										Left gyrus of Heschl
										Left Rolandic operculum
										Left postcentral gyrus
										Left supramarginal
2	204	0	71.38	5.64	−3	39	−15	Left rectus	11	Left rectus
		0	61.2	5.39	6	33	−15	Right rectus	11	Left orbitofrontal
										Right rectus
										Right orbitofrontal*
										Left anterior cingulate*
										Right anterior cingulate*
3	338	0	60.9	5.38	48	−21	48	Right postcentral gyrus	4	Right precentral gyrus
		0.001	39.45	4.68	66	−3	24	Right postcentral gyrus	43	Right postcentral gyrus
		0.001	36.88	4.57	42	−9	45	Right precentral gyrus	6	Right supramarginal
4	20	0	43.79	4.85	−27	−24	54	Left precentral†	4†	Left precentral
										Left postcentral
5	22	0.001	40.03	4.7	51	0	−36	Right inferior temporal	20	Right inferior temporal*
6	183	0.001	38.11	4.63	−6	−27	63	Left paracentral lobule	4	Left paracentral lobule
		0.001	33.63	4.43	15	−27	60	Right paracentral lobule	4	Right paracentral lobule
		0.003	27.77	4.12	15	−24	69	Right precentral	4	Right precentral
										Right supplementary motor area
										Right postcentral
										Left precuneus
										Right precentral
7	22	0.001	37.96	4.62	−3	12	−12	Left olfactory	25	Left olfactory
										Right olfactory
										Left caudate
										Right caudate*
8	77	0.001	33.37	4.41	69	−21	0	Right superior temporal	21	Right superior temporal
		0.002	33.05	4.4	60	−15	9	Right superior temporal	48	Right rolandic operculum
		0.002	31.79	4.34	51	−21	15	Right rolandic operculum	48	Right Heschl
										Right insula
										Right middle temporal*

Notes: Brodmann's areas and anatomic locations were determined using MRIcro and finding the closest reasonable value. k = number of voxels in cluster; Voxel p(FDR-cor) = false dicovery rate corrected p value of voxel; Voxel equivZ = equivalent z value for the voxel. *Region replicated from Kozel et al. (2004b). †Closest structure/BA.

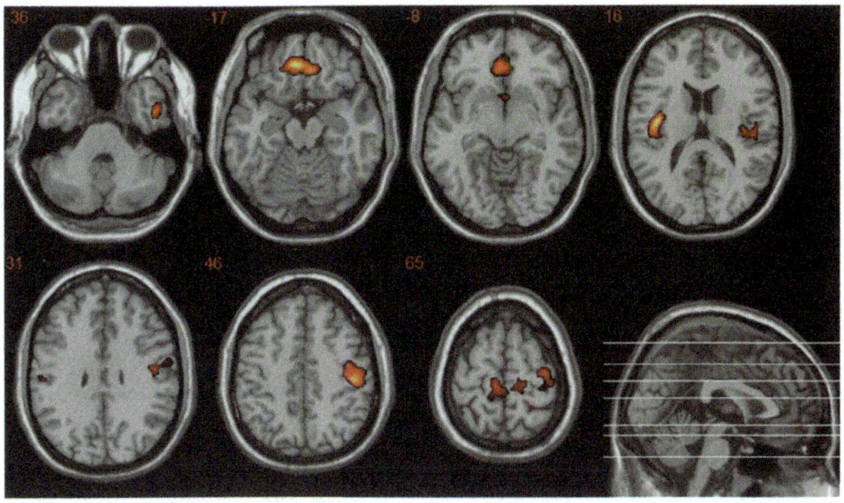

Figure 1. Group analysis of EDA correlated with BOLD fMRI signal: areas of significant activation (false discovery rate <0.005, k ≥20). The axial images of the structural template in MRIcro have areas of significant activation (FDR <0.005, k ≥20) overlaid in red to yellow. The sagittal image on the bottom right of the figure indicates the levels of the axial slices. The red numerals indicate the z level and their position signifies the left side of the image.

regressor and the number of participants whose calls were strengthened was about equal to the number whose calls were weakened. Thus, the acquiring and analyzing of simultaneous EDA data did not improve the diagnostic ability of our fMRI detection of deception model. This would suggest that EDA data (physiologic arousal) are not highly correlated with the measured brain imaging changes associated with our fMRI detection of deception methodology (i.e. the fMRI detection of deception methodology is not simply a more central measure of arousal).

The truthful and deceptive responses were not able to be differentiated by the EDA alone. This is not surprising given that the format of questioning was fundamentally different from the format that has been used in polygraph exams. In our paradigm, multiple questions are asked every 6 s versus a small number of questions being asked every 20–30 s in a polygraph exam. Also, we were employing only one of the four channels that polygraphers use to make their decisions.

Independent of task, significant associations were found in ventromedial and anterior cingulate brain regions. Although the MRI scanners (1.5Tesla Picker versus 3Tesla Philips), EDA acquisition equipment, and data analysis methods were different, certain brain regions replicated from our prior study (Kozel et al., 2004b). The right orbitofrontal and anterior cingulate regional activity again appears to correlate with EDA. These regions to varying degrees have been found by other investigators using various tasks and analysis methods (Critchley, Elliott, Mathias, & Dolan, 2000; Gamer, Bauermann, Stoeter, Vossel, 2007; Nagai, Critchley, Featherstone, Trimble, & Dolan, 2004; Patterson, Ungerleider, & Bandettini, 2002). Potential differences in results could be related to differences in task design, which aspect of the EDA signal was measured (value, change in height, etc.), and data analysis methodology.

A number of important factors need to be understood in order to properly interpret these data. The first is that the fMRI question format (a question every 6 s) is different than the typical polygraph format (20–30 s per question). If the fMRI question format was closer in timing to the polygraph format, the EDA might have significantly changed the results. Another important point is that the magnitude of risk in this laboratory experiment was relatively small ($50) compared to the greater stakes that would most likely be involved in real-world use. With greater risk, the importance of including the EDA values in the analysis may increase. Further testing will be needed to clarify these issues.

The lack of change in the fMRI detection of deception model accuracy and the lack of overlap between the brain regions associated with EDA and lie detection suggest that the fMRI model developed for deception and the polygraph are measuring unique aspects of deceptive behavior. Further work to investigate whether the two modalities could provide synergistic information in diagnosing deceptive behavior is needed.

In summary, a unique approach to analyze simultaneously collected EDA during an fMRI deception test was investigated. The addition of this measure did not improve our diagnostic ability. Further, we replicated the finding that EDA correlates with brain activity in the right orbitofrontal and bilateral anterior cingulate regions.

REFERENCES

Abe, N., Suzuki, M., Mori, E., Itoh, M., & Fujii, T. (2007). Deceiving others: Distinct neural responses of the prefrontal cortex and amygdala in simple fabrication and deception with social interactions. *Journal of Cognitive Neuroscience, 19*, 287–295.

Abe, N., Suzuki, M., Tsukiura, T., Mori, E., Yamaguchi, K., Itoh, M., et al. (2006). Dissociable roles of prefrontal and anterior cingulate cortices in deception. *Cerebral Cortex, 16*, 192–199.

Annett, M. (1970). A classification of hand preference by association analysis. *British Journal of Psychology, 61*, 303–321.

Ashburner, J., & Friston, K. J. (1999). Nonlinear spatial normalization using basis functions. *Human Brain Mapping, 7*, 254–266.

Critchley, H., Elliott, R., Mathias, C., & Dolan, R. (2000). Neural activity relating to generation and representation of galvanic skin response: A functional magnetic resonance imaging study. *Journal of Neuroscience, 20*, 3033–3040.

First, M., Spitzer, R., Williams, J., & Gibbon, M (1995). *Structured Clinical Interview for DSM-IV (SCID)*. Washington, DC: American Psychiatric Press.

Friston, K. (2004). Experimental design and statistical parametric mapping. In R. S. J. Frackowiak, K. J. Friston, C. D. Frith, R. J. Dolan., C. J. Price, S. Zeki, et al. (Eds.), *Human brain function* (2nd ed., pp. 599–634). Amsterdam: Elsevier Academic Press.

Furedy, J (1986). Lie detection as psychophysiological differentiation: Some fine lines. In M. Coles, E. Donchin, & S. Porges (Eds.), *Psychophysiology systems, processes and applications* (pp. 683–701). New York: Guilford Press.

Gamer, M., Bauermann, T., Stoeter, P., & Vossel, G. (2007). Covariations among fMRI, skin conductance, and behavioral data during processing of concealed information. *Human Brain Mapping, 28*, 1287–1301.

Ganis, G., Kosslyn, S. M., Stose, S., Thompson, W. L., & Yurgelun-Todd, D. A. (2003). Neural correlates of different types of deception: An fMRI investigation. *Cerebral Cortex, 13*, 830–836.

Geddes, L. A. (2002). The truth shall set you free. Development of the polygraph. *IEEE Engineering in Medicine and Biology Magazine, 21*, 97–100.

Johnson, K. A., George, M. S., & Kozel, F. A. (2007). Detecting deception using functional magnetic resonance imaging. *Directions in Psychiatry, 27*, in press.

Kozel, F. A., Johnson, K. A., Mu, Q., Grenesko, E. L., Laken, S. J., & George, M. S. (2005). Detecting deception using functional magnetic resonance imaging. *Biological Psychiatry, 58*, 605–613.

Kozel, F. A., Padgett, T. M., & George, M. S. (2004a). A replication study of the neural correlates of deception. *Behavioral Neuroscience, 118*, 852–856.

Kozel, F. A., Revell, L. J., Lorberbaum, J. P., Shastri, A., Elhai, J. D., Horner, M. D., et al. (2004b). A pilot study of functional magnetic resonance imaging brain correlates of deception in healthy young men. *Journal of Neuropsychiatry and Clinical Neuroscience, 16*, 295–305.

Langleben, D. D., Loughead, J. W., Bilker, W. B., Ruparel, K., Childress, A. R., Busch, S. I., et al. (2005). Telling truth from lie in individual subjects with fast event-related fMRI. *Human Brain Mapping, 26*, 262–272.

Langleben, D. D., Schroeder, L., Maldjian, J. A., Gur, R. C., McDonald, S., Ragland, J. D., et al. (2002). Brain activity during simulated deception: An event-related functional magnetic resonance study. *NeuroImage, 15*, 727–732.

Lee, T. M., Liu, H. L., Chan, C. C., Ng, Y. B., Fox, P. T., & Gao, J. H. (2005). Neural correlates of feigned memory impairment. *NeuroImage, 28*, 305–313.

Lee, T. M., Liu, H. L., Tan, L. H., Chan, C. C., Mahankali, S., Feng, C. M., et al. (2002). Lie detection by functional magnetic resonance imaging. *Human Brain Mapping, 15*, 157–164.

Mohamed, F. B., Faro, S. H., Gordon, N. J., Platek, S. M., Ahmad, H., & Williams, J. M. (2006). Brain mapping of deception and truth telling about an ecologically valid situation: Functional MR imaging and polygraph investigation—initial experience. *Radiology, 238*, 679–688.

Nagai, Y., Critchley, H. D., Featherstone, E., Trimble, M. R., & Dolan, R. J. (2004). Activity in ventromedial prefrontal cortex covaries with sympathetic skin conductance level: a physiological account of a "default mode" of brain function. *NeuroImage, 22*, 243–251.

Nunez, J. M., Casey, B. J., Egner, T., Hare, T., & Hirsch, J. (2005). Intentional false responding shares neural substrates with response conflict and cognitive control. *NeuroImage, 25*, 267–277.

Patterson, J. C., 2nd, Ungerleider, L. G., & Bandettini, P. A. (2002). Task-independent functional brain activity correlation with skin conductance changes: An fMRI study. *NeuroImage, 17*, 1797–1806.

Phan, K. L., Magalhaes, A., Ziemlewicz, T., Fitzgerald., D., Green, C., & Smith, W. (2005). Neural correlates of telling lies: A functional magnetic resonance imaging study at 4 tesla. *Academic Radiology, 12*, 164–172.

Spence, S. A., Farrow, T. F., Herford, A. E., Wilkinson, I. D., Zheng, Y., & Woodruff, P. W. (2001). Behavioural and functional anatomical correlates of deception in humans. *NeuroReport, 12*, 2849–2853.

SOCIAL NEUROSCIENCE, 2009, 4 (6), 518–527

Psychology Press
Taylor & Francis Group

Towards clinical trials of lie detection with fMRI

J. G. Hakun, K. Ruparel, D. Seelig, E. Busch, J. W. Loughead, R. C. Gur, and D. D. Langleben

University of Pennsylvania, Philadelphia, Pennsylvania, USA

Recent reports of successful fMRI-based discrimination between lie and truth in single subjects raised the interest of prospective users and a public concern about the potential scope of this technology. The increased scrutiny highlighted the lack of controlled "real life", i.e. prospective clinical trials of this technology that conform to the common standards of medical device development. The ethics of conducting such trials given the paucity of data on fMRI-based lie detection has also been questioned. To probe the potential issues of translating the laboratory research into practice, we conducted a case study in which we adapted the standard Guilty Knowledge Test (GKT), a well-established model of producing deception, to the common scenario of lying on a resumé. The task consisted of questions about pertinent items on the subject's resumé, three of which could be independently verified as truth (KNOWN) and three that could not be verified and were thus termed UNKNOWN. The subject had an incentive to lie on all UNKNOWN items, and on debriefing confirmed that he had done so. Data was preprocessed, masked with *a priori* regions of interest, thresholded, and qualitatively evaluated for consistency with the previously reported prefronto-parietal Lie > Truth pattern. Deceptive responses to two out of the three UNKNOWN items were associated with the predicted prefronto-parietal fMRI pattern. In the third UNKNOWN this pattern was absent, and instead, increased limbic (amygdala and hippocampus) response was observed. Based on published prefronto-parietal Lie response pattern, only the first two items could be categorized as Lie. If confirmed, this demonstration of amygdala and hippocampus activation in a Lie > Truth contrast illustrates the need to integrate the limbic system and its emotional and cognitive correlates into the existing model of deception. Our experiment suggests an approach to a naturalistic scenario and the research questions that need to be answered in order to set the stage for prospective clinical trials of fMRI-based lie detection.

INTRODUCTION

Deceit is ubiquitous in humans, and almost every generation has attempted to apply technology of the age to understanding and detecting it (Eck, 1970). The advent of functional magnetic resonance imaging (fMRI), a reliable and safe measure of regional brain activity, inevitably led to attempts to utilize it for this purpose (reviewed in Spence, 2004). The first generation of fMRI studies of deception was dedicated to the initial demonstration of feasibility (Langleben et al., 2005; Lee et al., 2002; Spence et al., 2001). Subsequent work diverged into two arms: one devoted to further refinement of lie detection techniques and the other principally concerned with functional neuroanatomy and cognition (Davatzikos et al., 2005; Ganis, Kosslyn, Stose, Thompson, & Yurgelun-Todd, 2003; Kozel, 2004, 2005; Nunez, Casey, Egner, Hare, & Hirsch, 2005). Paradigms advanced from group average to single subject studies, demonstrating the potential contributions of memory, learning, emotion, and complex cognition to brain activity

Correspondence should be addressed to: Daniel D. Langleben, Treatment Research Center, University of Pennsylvania, 3900 Chestnut St., Philadelphia, PA 19104-6178, USA. E-mail: langlebe@mail.med.upenn.edu

This work was supported in part by a grant from the US Army Office of Research (ARO). Parts of this report appeared in *The Washington Post* (Garreau, 2006).

www.psypress.com/socialneuroscience DOI:10.1080/17470910802188370

patterns of deception (Abe, Suzuki, Mori, Itoh, & Fujii, 2007; Ganis et al., 2003; Nunez et al., 2005). Task-related features that covary with deception, such as stimulus familiarity, salience, motor response switching, and oddball interference, were also examined.

There is a converging model of deception as a working memory-intensive task, mediated to a large extent by the prefronto-parietal systems dedicated to behavioral control and attention (Langleben et al., 2002; Leung, Selig, & Gore, 2004; Nunez et al., 2005; Spence et al., 2004). Recent experiments began refining this model by demonstrating the potential contributions of the limbic system to deception, as was anticipated in some earlier theoretical work (Abe et al., 2007; Langleben, Dattilio, & Guthei, 2006). The high degree of reproducibility and accuracy in single subjects raised academic and public curiosity, commercial activity, and political concerns (Henig, 2006; Talbot, 2007). The more fervent critics of the new technology assailed it for being potentially effective enough to warrant its ban to protect "cognitive freedoms." Others had more realistic concerns with its relatively low specificity, which would translate to false positive rates not acceptable in legal applications (Halber, 2007) or for employment screening (Stern, 2004).

Yet another criticism was that it is impossible to conceive of ethically sound prospective clinical trials of fMRI-based lie-detection and equally impossible to proceed to application of this technique to real-life cases without data acquired in such trials (Kanwisher, quoted in Halber (2007)). This debate highlights the lack of critical translational data (Wolpe, Foster, & Langleben, 2005) that is necessary to determine the true potential and limitations of fMRI in lie-detection.

Designing clinical trials of fMRI-based lie detection implies using unproven technology in situations where balancing the risk/benefit ratio represents a legal and ethical challenge (Appelbaum, 2007; Thompson, 2005). To cautiously probe the possibility that this challenge is addressable, we present a case of a volunteer experimental participant (JG) who was writing a story for the *Washington Post* and wanted to test the fMRI method on himself. He intentionally simulated overstatements of personal accomplishments on a resumé, as one would for the purpose of gaining undeserved employment at a fictional job. This experiment closely resembles the recent events involving the Dean of Admissions at the Massachusetts Institute of Technology (Lewin, 2007). The fact that our experiment took place almost a year prior to this scandal may highlight its ecological validity.

Testing resumé statements presented one challenge that had not yet been reported in fMRI studies of deception: the presence of multiple items of unknown veracity, thus limiting our ability to contrast Lie and Truth and demanding that we contrast response to each unknown item with known Truth instead. The experimental paradigm designed to address this case was termed the "Resumé Query Test" (RQT). Our working hypotheses were: (1) The pattern of brain response to Lie during RQT will be similar to the prefronto-parietal pattern we observed in prior forced-choice deception paradigms (Langleben, 2005); (2) Lie and Truth could be distinguished by qualitative inspection of the relevant contrasts, thresholded at $p < 0.05$, and masked with a region of interest (ROI) template (see Methods).

METHODS

Participant

JG is a 53-year-old male, healthy, right-handed newspaper reporter. JG gave written consent to publish this manuscript with identifying information included. The study protocol was approved by the University of Pennsylvania IRB.

fMRI paradigm: "resumé query" task (RQT)

Scenario

The subject was instructed by one of the authors (DDL) to pretend that he was applying for a job as a science writer, with DDL acting as a recruiter and a co-conspirator and another author (JWL) as a representative of the hiring manager. DDL reviewed JG's resumé and informed him that the key requirements for the job were military service, a college degree, and writing experience that should include two published books. DDL, highlighting the key requirements, then returned the resumé to JG and told him to resubmit it, and accompany these requirements with supporting documents regarding his military service, college degree and writing experience (which were to be used as task stimuli for the fMRI exam). JG was instructed to "employ whatever means necessary to be hired," which provided him with a strong incentive to falsify one or more of those items if they were missing from his resumé. The other members of the research team were blinded to all details of this transaction. JG was also told to assume that successful hiring would require that he successfully passed an fMRI exam designed to probe the veracity of some of the items listed on his resumé. Other than the instruction to "employ whatever means necessary" to be hired, the subject was not explicitly instructed to falsify his resumé information. After receiving the subject's (modified) resumé, DDL selected three items that were discrepant with the original resumé (i.e. likely to be a lie) and three statements that could be easily verified as truthful. On the day of the imaging part of the experiment, the subject was escorted to the scanner by a third party and greeted by the MRI examiner (JWL), who informed him that he would test him with fMRI concerning the veracity of his resumé and instructed him to answer each item as accurately and truthfully as possible.

Task design

The subject was instructed to respond "Yes" or "No" to the questions regarding the six specific resumé statements, accompanied by appropriate images of supporting materials he provided with his resumé (Figure 1). The six questions were divided equally between two stimulus classes: (1) "Known truths" (KNOWN)—questions designed to be easily verifiable as truths, and (2) "Unknowns" (UNKNOWN)—questions relevant to job qualification and difficult to verify by a hypothetical job interviewer. The three KNOWN questions probed in this task were: (1) "Did you write this book?"—accompanied by an image of the cover of a book written by the subject; (2) "Do you work for the *Post*?"—accompanied by an image of the cover of the *Washington Post* with an article written by the subject; and (3) "Is this your home?"—accompanied by a picture of the subject's house. The three UNKNOWN questions probed in this task were: (1) "Did you write this book?"—accompanied by an image of the cover of a book not written by the subject, but altered to feature the subject's name as the author; (2) "Did you graduate from Notre Dame?"—accompanied by an image of a Notre Dame University BA diploma with the subject's name as the graduate; and (3) "Did you serve in the military?"—accompanied by an image of a certificate of honorable discharge from the US Marine Corps (USMC), altered to feature the subject's name and a fictional service number.

Each question was repeated 15 times, in a pseudorandom order optimized for event-related fMRI (Dale, 1999). Questions were presented for

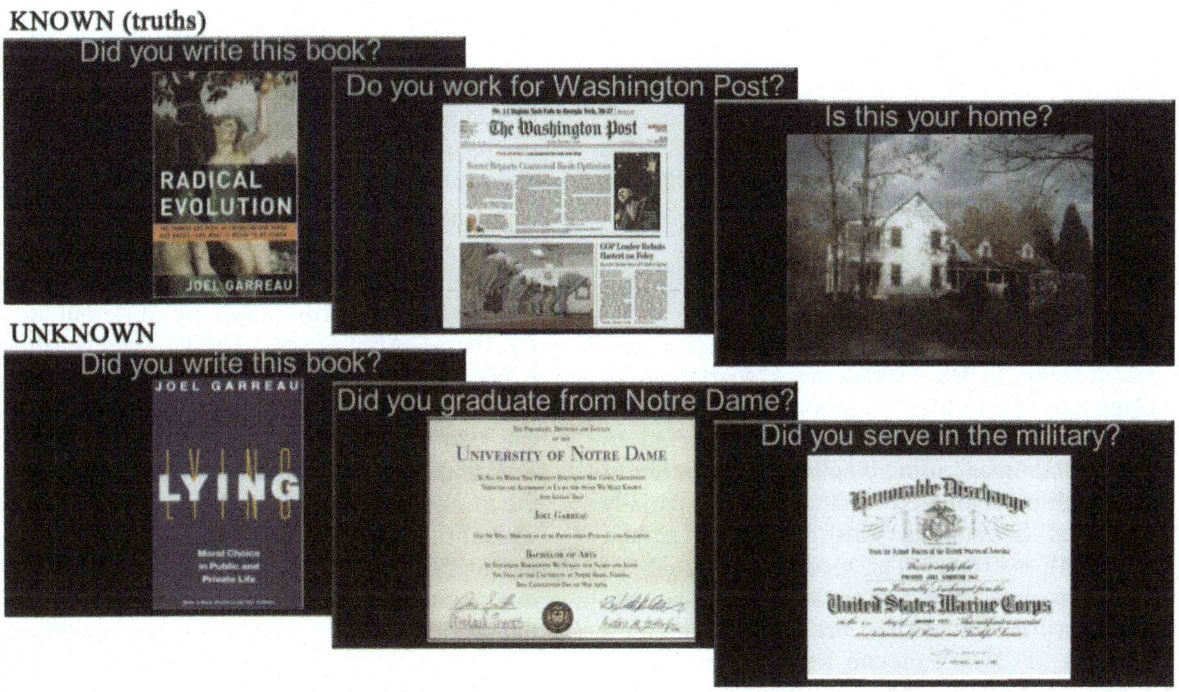

Figure 1. RQT stimuli: Known and unknown RQT questions.

2 s followed by a variable "jittered" interstimulus interval (ISI; 1–13 s, average 2 s). During the interval, a yellow asterisk was displayed in the center of the black screen (NULL event). A total of 90 questions were presented, for a total task length of 9 min.

Stimuli were rear-projected to the center of the visual field using a video projector (PowerLite 7300; Epson America, Long Beach, CA) viewed through a head-coil-mounted mirror. Stimuli presentation was synchronized with image acquisition using Presentation (version 11.2, www.neurobs.com). Responses were made with a right (dominant) hand press on a two-button fiberoptic-response keypad (fORP; Current Design, Philadelphia, PA) and logged by Presentation software.

The participant was debriefed about the way he responded and queried as to which UNKNOWN items he selected to lie about after the analysis was complete and available.

Image acquisition

Images were acquired with blood-oxygen level dependent (BOLD) imaging (Bandettini, Wong, Hinks, Tikofsky, & Hyde, 1992) on a clinical 3 T Siemens Trio Scanner (Iselin, NJ). A 5-min, magnetization prepared, rapid-acquisition gradient echo image (MPRAGE) was acquired for anatomic overlays of functional data and spatial normalization (T1 MNI Template). BOLD imaging used a 33-slice whole-brain, single-shot gradient echo (GE) echo-planar (EPI) sequence (TR/TE = 3000/21 ms, FOV = 240 mm, matrix = 64×64, slice thickness/gap = 4/0 mm). This sequence delivered a nominal voxel resolution of $3 \times 3 \times 3$ mm.

Data analysis

The fMRI data were preprocessed and analyzed using SPM2 (Wellcome Department of Cognitive Neurology, London, UK). Images were slice time-corrected, motion-corrected to the median image using b-spline interpolation (4 df), high-pass filtered (100 s), and spatially smoothed (8 mm full width at half maximum (FWHM), isotropic). The median functional volume was coregistered to the anatomical volume and then transformed into the standard anatomical space (T1 MNI template, resampled to $2 \times 2 \times 2$ mm resolution) using trilinear interpolation (Ashburner & Friston, 1999). Statistical analysis used a general linear model (GLM) that incorporated an event-related design, which contained a total of four regressors along with their temporal derivatives and an intercept term. Regressors were created by convolving each of the stimuli timecourses with the standard double gamma hemodynamic response function (HRF). The three KNOWN class stimuli were concatenated and entered into the GLM as one regressor. Each of the three UNKNOWN class stimuli was entered as a separate regressor. Statistical contrast activation maps were generated by subtracting the regressor weights (beta-coefficients) associated with the conditions of interest (e.g. UNKNOWN > KNOWN and KNOWN > UNKNOWN) in a random effects model.

Each UKNOWN > KNOWN was masked with an anatomical ROI mask derived from the Wake Forest University PickAtlas (Maldjian, Laurienti, Kraft, & Burdette, 2003) and the regions previously identified as important to deception (Abe et al., 2007; Kozel et al., 2005; Langleben et al., 2005). The mask included both predicted areas associated with Lie > Truth and parts of the limbic system as part of an exploratory analysis. The ROI mask included Broadman's areas (BA) 13, 22, 39, 40, 41, 45, 47; dilated by 2 mm. Broadman's areas 13, 45, 47 (inferior frontal gyrus, IFG), 40 (inferior parietal lobe, IPL and supramarginal gyrus, SMG), 22, 41 (superior temporal gyrus, STG), 39 (middle temporal gyrus, MTG), and 11 (Medial Frontal Gyrus, MFG) were chosen to be part of the *a priori* lie-template mask due to their implication in previous studies as regions associated with Lie > Truth. BA 28 (hippocampus, Hipp); 34 (amygdala, Amyg and uncus); and 36 (parahippocampal gyrus, PHG) were included due to their association with the limbic system and recent report of limbic activation during deception where lying was both endorsed and unendorsed by different team members (Abe et al., 2007). Average of all KNOWN > average of all UNKNOWN contrast was left unmasked as the analysis of this contrast was also exploratory and not hypothesis-driven.

Monte Carlo simulations were performed on the masked and thresholded dataset to determine a dual-threshold (voxel probability and cluster extent probability) family-wise error correction rate using AFNI's Alphasim (Ward, 2000). With this method it was determined that a voxel-height probability of $p < .05$ and a minimum cluster extent of 83 voxels would achieve an $\alpha < .05$ for this masked dataset. All result images were examined at two thresholds: (1) $p < .05$ family-wise corrected for multiple comparisons; and (2) $p < .05$ (minimum 20 contiguous significant voxels) uncorrected for multiple comparisons for exploratory analysis of clusters below the family-wise threshold. This dual thresholding of result images controls for Type I errors, but also allows for further investigation of subthreshold results in this single-subject example.

RESULTS

Debriefing

JG confirmed that the three KNOWNS were Truth and indicated that all three UNKNOWNS were Lie and that he lied about all three during the task. JG's claim that he lied about all three items throughout the task was corroborated by analysis of the behavioral response data.

RQT task fMRI results

In the contrast of UKNOWN 1 with the average of KNOWN truth, UNKNOWN 1: "Did you write this book?" revealed significant bilateral activation in IFG (BA 13, 47) and IPL/SMG (BA 40) (Table 1). In addition, there was activation in bilateral STG (BA 22, 41) and right PHG (BA 36). In terms of height and extent of activation, the most prominent activations in UKNOWN 1 were in the left IFG and the right SMG, though both IFG and SMG activations were bilateral.

When UNKNOWN 2: "Did you graduate from Notre Dame?" was contrasted with the average of KNOWN truth, the contrast revealed significant activation in bilateral IFG (BA 45, 47) and IPL/SMG (BA 40). In addition, activation was observed in right subcallosal gyrus, SCG (BA 25), medial frontal gyrus (BA 11), bilateral PHG (BA 35, 36), and left uncus (BA 34). Activations in left and right IPL were equally significant in terms of height and extent; however, left IFG activation was more significant than right IFG activation in both height and extent.

When UKNOWN 3: "Did you serve in the Military?" was contrasted with the average of KNOWN truth, the contrast revealed significant activation in right hippocampus (BA 28) and MTG (BA 39). Also, activation was observed in the right amygdala (BA 34) and left IPL/SMG (BA 40).

Lastly, when the average of all KNOWN was contrasted with the average of all UNKNOWN, the contrast revealed significant activation in the right MFG (BA 11), left MTG (BA 21, 39), and right inferior temporal gyrus, ITG (BA 37) (Figure 2).

DISCUSSION

Three members of the study team, blinded as to which, if any, of the UNKNOWN items were lies, evaluated the UNKNOWN > KNOWN contrasts and provided one consensus opinion. Deceptive responses in two of the three UNKNOWN items were associated with increased prefronto-parietal activation (IFG, BA 45, 47 and IPL/SMG, BA 40), as predicted in our first hypothesis, and could be categorized as Lie, as predicted in our second hypothesis. The inferior frontal and inferior parietal are the two regions consistently associated with deception in Lie > Truth comparisons under the GKT model, especially after salience is controlled for (Gamer, Bauermann, Stoeter, & Vossel, 2007; Langleben et al., 2005). The RQT critically departs from the laboratory GKT model by the presence of multiple unknowns instead of one endorsed or possible lie. Similarly to GKT1, the RQT controls for the prepotency of response type, i.e. all responses were uniformly "Yes" (RQT) or uniformly "No" (GKT1), which changing can in itself activate the prefronto-parietal network (Linden et al., 1999). Also similarly to GKT2, the RQT controls for the familiarity of task items. Thus, the RQT carries enough similarities with the GKT models (Elaad, Ginton, & Jungman, 1992; Langleben et al., 2005) to consider it an adaptation of the GKT1 and GKT2 to a real-life scenario rather than an entirely new task. Therefore, similarities in activation pattern in the RQT and GKT during Lie are expected.

The third UNKNOWN contrast (Figure 2, "Did you serve in the military?") stands out. Contrary to our expectations, it was associated with amygdala and hippocampus activation and, while there was activation in the left IPL, there was no prefrontal response. Thus, an expert using a prefronto-parietal model of Lie > Truth brain activation may be unable to categorize this response as a Lie. This may be the first report of limbic activation during Lie in a forced-choice unendorsed deception paradigm, though Abe et al. (2007) found amygdala activation as a

TABLE 1
fMRI results of each UNKNOWN >ALL KNOWN

REGION	*Unknown 1: Book*						*Unknown 2: Notre Dame*						*Unknown 3: Military*					
Frontal	*No. vox*	*BA*	*X*	*Y*	*Z*	Z_{max}	*No. vox*	*BA*	*X*	*Y*	*Z*	Z_{max}	*No. vox*	*BA*	*X*	*Y*	*Z*	Z_{max}
IFG L	**531**	**13,47**	**−38**	**12**	**−8**	**3.16**	**900**	**47**	**−36**	**17**	**−4**	**3.65**						
	209	**45**	**−48**	**20**	**19**	**3.21**	**190**	**45**	**−45**	**36**	**2**	**2.99**						
	61	45	−46	34	6	2.44												
IFG R	37	47	44	17	−10	2.29	**1038**	**47**	**32**	**21**	**−4**	**3.26**						
	29	45	43	19	18	2.29	**241**	**45**	**−45**	**16**	**18**	**3.1**						
SCG R							27	25	11	14	−11	2.68						
MFG L							29	11	−14	30	−14	2.66						
Temporal																		
STG L	71	22	−58	−43	20	2.63												
STG R	33	41	42	−34	14	2.46												
MTG R													31	39	46	−65	26	2.17
PHG L							**100**	**35**	**−23**	**−19**	**−16**	**3.25**						
							71	36	−24	−36	−11	2.92						
PHG R	24	36	24	−30	−15	2.55	**246**	**35**	**21**	**−30**	**−12**	**2.83**						
							40	36	24	−21	−24	2.29						
Hipp R													**142**	**28**	**24**	**−18**	**−19**	**3.13**
Amyg R													40	34	20	−3	−14	2
Uncus L							**188**	**34**	**−24**	**3**	**−19**	**3.38**						
Parietal																		
IPL/SMG L	**149**	**40**	**−43**	**−40**	**38**	**2.67**	**1937**	**40**	**−44**	**−37**	**34**	**3.89**	45	40	−42	−38	36	2.16
	62	40	−32	−60	41	2.89												
IPL/SMG R	**1079**	**40**	**42**	**−51**	**35**	**3.66**	**1900**	**40**	**42**	**−46**	**32**	**3.98**						

Notes: Data in table is thresholded at $p < .05$, uncorrected for multiple comparisons; clusters of activation that survive family-wise correction at $p < .05$ are given in bold type. UNKNOWN 1: "Did you write this book?", UNKNOWN 2: "Did you graduate from Notre Dame?", UNKNOWN 3: "Did you serve in the military?", UNKNOWNs 1 and 2 categorized as lies due to presence of significant activation in IFG and SMG; UNKNOWN 3 categorized as a truth due to lack of IFG activation. No. of voxels expressed as number of contiguous voxels within ROI. Z_{max} refers to the peak Z within a given ROI. Empty rows denote no super-threshold voxels within ROI.

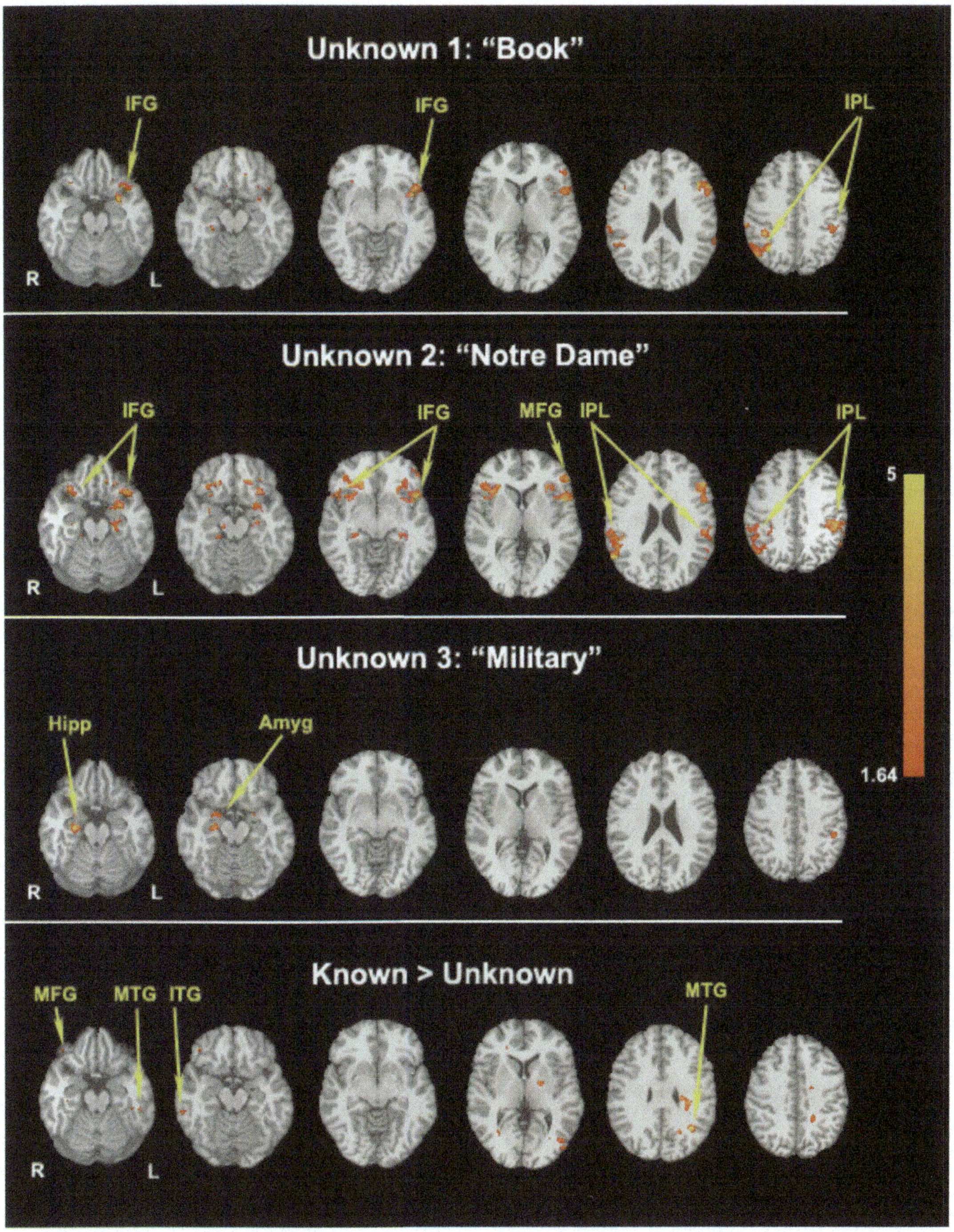

Figure 2. UNKNOWN > KNOWN and KNOWN > UNKNOWN fMRI results. fMRI results of each UNKNOWN > KNOWN, and average of KNOWN > average of UNKNOWN. Top and second row: UNKNOWN 1 and 2 exhibit heightened response in the bilateral IFG and IPL/SMG. UNKNOWN 3 (third row) exhibited heightened activation in the left IPL, right hippocampus and amygdala. Average of KNOWN > average of UNKNOWN (bottom row) exhibited heightened response in the MFG, ITG, and MTG. Single subject results registered to and projected over the MNI template and masked with the *a priori* ROI mask (see Methods). Result images displayed at $p < 0.05$ uncorrected for multiple comparisons.

main effect of non-endorsed responses when a subject engaged in both lying AND truth-telling.

The limbic activation could be due to either arousal related to the topic of the USMC, or specifically to lying about this particular question. The former may relate to JG's feelings about the military that were formed while he was an "embedded" reporter with the USMC (Garreau,

2006). The second may relate to the increased limbic response associated with social inappropriateness (Finger, Marsh, Kamel, Mitchell, & Blair, 2006) and moral dilemmas (Greene, Nystrom, Engell, Darley, & Cohen, 2004), features that are common to the RQT format and may be inherent to lying about socially or personally significant questions.

The concurrent absence of prefronto-parietal response in UNKNOWN 3 with increase in limbic response can be interpreted in several ways. The limbic system could inhibit activation of the prefrontal system mediating deception under conditions of increased emotional and memory involvement (Shafritz, Collins, & Blumberg, 2006). It is also possible that JG did not subjectively perceive an affirmative response to this question as a lie, given his experience of being "embedded" with the USMC, a cognitive strategy that JG admitted employing as a countermeasure during debriefing (Garreau, 2006). Finally, the greater difficulty in matching the features of the stimuli in the RQT (in contrast to the GKT, where all items were playing cards) may have confounded the anticipated lie response in IFG and IPL for UNKNOWN 3. Analyzing these data in the style of a controlled question task (CQT) would require proper matching of each UNKNOWN question with a well-matched control question, and might help reduce or eliminate false negative results. This possibility underscores the importance of considering the perceptual features of naturalistic stimuli in future applications of the RQT. Lastly, in order to help interpret counterintuitive results such as those seen in UNKNOWN 3, it would be useful to assess, for each query item, participants' subjective emotions about lying.

Though the findings in this report cannot be simply generalized to populations, they illustrate a number of issues in the neurobiology of deception that should be investigated before conducting clinical trials. First, future studies using scenarios with higher ecological validity may discover fMRI patterns associated with deception, other than those reported so far. Specifically, the absence of limbic system activation in Lie contrasts in all prior fMRI studies is puzzling, considering the role of the limbic system in emotion, memory, and anxiety. This may be due to the technical difficulties in acquiring fMRI signal from the inferior part of the brain or to the low risk/benefit ratio of deception in the laboratory setting. Our RQT paradigm is example of a transitional model, midway between the laboratory and real life, that could provide data to justify prospective clinical trials of fMRI-based lie detection in relevant populations. Elucidating the relationship between the limbic system and the prefronto-parietal network may be one of the missions of such transitional experiments. Second, a major practical limitation of applying the current GKT paradigms to real life is the inability to fully control the visual salience of query items. For example, while the real and forged book covers were well matched, the visual comparability of a forged diploma to an image of one's home is less clear. This issue may be more easily addressed in the laboratory than in the field. One solution is to derive empirical corrections for features such as complexity and lighting, another is to use text-only query items. Such approaches have not been studied and require validation.

The list of the unaddressed issues en route to conclusive clinical trials could be longer. One such issue is identifying a consensus approach for reporting clinical single subject fMRI data. This issue has only recently begun to be elaborated upon in the literature on deception (Kozel, 2008). Specifically means for adequate control of Type I errors is required. Though anatomical masking may be an effective in reducing multiple comparisons, it precludes investigation of unpredicted task-related activation, a potential problem in a situation that may involve a high level of individual variability.

A more theoretical but no less important issue is the concept of theory of mind (TOM) and its place in the neuroscience of deception (Stuss, 2001). Tests that elicit deception for the purpose of detection, including the RQT, do not address TOM, which is invariably engaged by a deceptive social interaction. In order to engage TOM, our RQT could be modified to emulate a human interviewer.

Finally, we would like to propose a distinction between lie-detection and "mind-reading." Lying involves a conscious act of suppression or modification of subjective truth in response to a query. Identifying deception by measuring its physiological markers is the aim of lie-detection. On the other hand, "mind-reading" implies making inference about the cognitive or emotional state of an individual through pattern-recognition of fMRI data without explicit query or conscious behavioral response (Haynes & Rees, 2006; Haynes et al., 2007; Norman, Polyn, Detre, & Haxby, 2006). While potentially very powerful, this approach is currently more vulnerable to interpretation than a forced-choice lie-detection paradigm involving predefined query such as the RQT.

This case report illustrates an approach and the hurdles therein on the way from laboratory experiments to prospective clinical trials of fMRI-based lie detection. Our findings suggest that under conditions closely approximating a real-life situation, fMRI signal patterns associated with deception generated by a GKT-type model may be more variable than previously reported. Our observations indicate that increasing ecological validity of the deception task may not necessarily diminish the accuracy of lie detection, but it may require additional studies to address variations associated with the social and personal consequences of deception scenarios. The study also illustrates the need to further investigate the contribution of emotion and memory to brain activation evoked by deception scenarios with high ecological validity. Finally, our case provides an example of an ethically sound experiment that bridges the gap between the laboratory and clinical trial.

REFERENCES

Abe, N., Suzuki, M., Mori, E., Itoh, M., & Fujii, T. (2007). Deceiving others: Distinct neural responses of the prefrontal cortex and amygdala in simple fabrication and deception with social interactions. *Journal of Cognitive Neuroscience, 19*(2), 287–295.

Appelbaum, P. S. (2007). Law & psychiatry: The new lie detectors: Neuroscience, deception, and the courts. *Psychiatric Services, 58*(4), 460–462.

Ashburner, J., & Friston, K. J. (1999). Image registration. In C. T. W. Moonen & P. A. Bandettini (Eds.), *Functional MRI* (pp. 285–299). New York: Springer.

Bandettini, P. A., Wong, E. C., Hinks, R. S., Tikofsky, R. S., & Hyde, J. S. (1992). Time course EPI of human brain function during task activation. *Magnetic Resonance in Medicine, 25*(2), 390–397.

Dale, A. M. (1999). Optimal experimental design for event-related fMRI. *Human Brain Mapping, 8*(2–3), 109–114.

Davatzikos, C., Ruparel, K., Fan, Y., Shen, D. G., Acharyya, M., Loughead, J. W., Gur, R. C., & Langleben, D. D. (2005). Classifying spatial patterns of brain activity with machine learning methods: Application to lie detection. *NeuroImage, 28*(3), 663–668.

Eck, M. (1970). *Lies and truth* (B. Murchland, Trans.). New York: Macmillan.

Elaad, E., Ginton, A., & Jungman, N. (1992). Detection measures in real-life criminal guilty knowledge tests. *Journal of Applied Psychology, 77*(5), 757–767.

Finger, E. C., Marsh, A. A., Kamel, N., Mitchell, D. G., & Blair, J. R. (2006). Caught in the act: The impact of audience on the neural response to morally and socially inappropriate behavior. *NeuroImage, 33*(1), 414–421.

Gamer, M., Bauermann, T., Stoeter, P., & Vossel, G. (2007). Covariations among fMRI, skin conductance, and behavioral data during processing of concealed information. *Human Brain Mapping, 28*, 1287–1301.

Ganis, G., Kosslyn, S. M., Stose, S., Thompson, W. L., & Yurgelun-Todd, D. A. (2003). Neural correlates of different types of deception: An FMRI investigation. *Cerebral Cortex, 13*(8), 830–836.

Garreau, J. (2006, October 30). Brain on fire: Tell a whopper and watch the screen light up: Thanks (or no thanks) to sophisticated scanning, the lie may be on its last legs. *Washington Post*, pp. C1–C2.

Greene, J. D., Nystrom, L. E., Engell, A. D., Darley, J. M., & Cohen, J. D. (2004). The neural bases of cognitive conflict and control in moral judgment. *Neuron, 44*(2), 389–400.

Halber, D. (2007, February). *Scientists swear a good lie detector is hard to find.* Paper presented at the MIT Tech Talk, Cambridge, MA.

Haynes, J. D., & Rees, G. (2006). Decoding mental states from brain activity in humans. *Nature Reviews Neuroscience, 7*(7), 523–534.

Haynes, J. D., Sakai, K., Rees, G., Gilbert, S., Frith, C., & Passingham, R. E. (2007). Reading hidden intentions in the human brain. *Current Biology, 17*(4), 323–328.

Henig, R. M. (2006, February 5). Looking for the lie: Scientists are using brain imaging and other tools as new kinds of lie detectors. But trickier even than finding the source of deception might be navigating a world without it. *New York Times Magazine*, 47–53, 76, 80.

Kozel, F. A., Johnson, K. A., Mu, Q., Grenesko, E. L., Laken, S. J., & George, M. S. (2005). Detecting deception using functional magnetic resonance imaging. *Biological Psychiatry, 58*(8), 605–613.

Kozel, F. A., & Madhukar, H. T. (2008). Developing a neuropsychiatric functional brain imaging test. *Neuroscience and crime (special issue of* Neurocase). Hove, UK: Psychology Press.

Kozel, F. A., Revell, L. J., Lorberbaum, J. P., Shastri, A., Elhai, J. D., Horner, M. D., Smith, A., Nahas, Z., Bohning, D. E., & George, M. S. (2004). A pilot study of functional magnetic resonance imaging brain correlates of deception in healthy young men. *Journal of Neuropsychiatry and Clinical Neurosciences, 16*, 295–305.

Langleben, D. D., Dattilio, F. M., & Guthei, T. G. (2006). True lies: Delusions and lie-detection technology. *The Journal of Psychiatry & Law, 34*(Fall), 351–370.

Langleben, D. D., Loughead, J. W., Bilker, W. B., Ruparel, K., Childress, A. R., Busch, S. I., & Gur, R. C. (2005). Telling truth from lie in individual subjects with fast event-related fMRI. *Human Brain Mapping, 26*(4), 262–272.

Langleben, D. D., Schroeder, L., Maldjian, J. A., Gur, R. C., McDonald, S., Ragland, J. D., O'Brien, C. P., &

Childress, A. R. (2002). Brain activity during simulated deception: An event-related functional magnetic resonance study. *NeuroImage, 15*(3), 727–732.

Lee, T. M., Liu, H. L., Tan, L. H., Chan, C. C., Mahankali, S., Feng, C. M., Hou, J., Fox, P. T., & Gao, J. H. (2002). Lie detection by functional magnetic resonance imaging. *Human Brain Mapping, 15*(3), 157–164.

Leung, H. C., Seelig, D., & Gore, J. C. (2004). The effect of memory load on cortical activity in the spatial working memory circuit. *Cognitive, Affective and Behavioral Neuroscience, 4*(4), 553–563.

Lewin, T. (2007, April 26). Dean at M.I.T. resigns, ending a 28-year lie [Electronic version]. *The New York Times.* Retrieved May 13, 2008, from www.nytimes.com

Linden, D. E., Prvulovic, D., Formisano, E., Vollinger, M., Zanella, F. E., Goebel, R., & Dierks, T. (1999). The functional neuroanatomy of target detection: An fMRI study of visual and auditory oddball tasks. *Cerebral Cortex, 9*(8), 815–823.

Maldjian, J. A., Laurienti, P. J., Kraft, R. A., & Burdette, J. H. (2003). An automated method for neuroanatomic and cytoarchitectonic atlas-based interrogation of fMRI data sets. *NeuroImage, 19*(3), 1233–1239.

Norman, K. A., Polyn, S. M., Detre, G. J., & Haxby, J. V. (2006). Beyond mind-reading: Multi-voxel pattern analysis of fMRI data. *Trends in Cognitive Sciences, 10*(9), 424–430.

Nunez, J. M., Casey, B. J., Egner, T., Hare, T., & Hirsch, J. (2005). Intentional false responding shares neural substrates with response conflict and cognitive control. *NeuroImage, 25*(1), 267–277.

Shafritz, K. M., Collins, S. H., & Blumberg, H. P. (2006). The interaction of emotional and cognitive neural systems in emotionally guided response inhibition. *NeuroImage, 31*(1), 468–475.

Spence, S. A. (2004). The deceptive brain. *Journal of the Royal Society of Medicine, 97*(1), 6–9.

Spence, S. A., Farrow, T. F., Herford, A. E., Wilkinson, I. D., Zheng, Y., & Woodruff, P. W. (2001). Behavioral and functional anatomical correlates of deception in humans. *NeuroReport, 12*(13), 2849–2853.

Stern, P. C. (Ed.). (2004). *The polygraph and lie detection* (1st ed.). Report of The National Research Council Committee to Review the Scientific Evidence on the Polygraph. Washington, DC: The National Academies Press.

Stuss, D. T., Gallup, G. G., Jr., & Alexander, M. P. (2001). The frontal lobes are necessary for 'theory of mind'. *Brain, 124*(2), 279–286.

Talbot, M. (2007, July 2). Duped: Can brain scans uncover lies? *New Yorker*, 52–61.

Thompson, S. K. (2005). The legality of the use of psychiatric neuroimaging in intelligence interrogation. *Cornell Law Review, 90*(6), 1601–1637.

Ward, B. D. (2000). *Simultaneous inference for FMRI data.* Milwaukee, WI: Medical College of Wisconsin.

Wolpe, P. R., Foster, K. R., & Langleben, D. D. (2005). Emerging neurotechnologies for lie-detection: Promises and perils. *American Journal of Bioethics, 5*(2), 39–49.

SOCIAL NEUROSCIENCE, 2009, 4 (6), 528–538

Psychology Press
Taylor & Francis Group

Detection of deception using fMRI: Better than chance, but well below perfection

George T. Monteleone, K. Luan Phan, Howard C. Nusbaum, Daniel Fitzgerald, and John-Stockton Irick
University of Chicago, Chicago, IL, USA

Stephen E. Fienberg
Carnegie-Mellon University, Pittsburgh, PA, USA

John T. Cacioppo
University of Chicago, Chicago, IL, USA

Functional brain imaging has been considered a new and better technique for the detection of deception. The reasoning is that there is a neural locus or circuit for lying that is sensitive, specific, generalizable across individuals and measurement contexts, and robust to countermeasures. To determine the extent to which the group results predicted lying at the level of the individual, we reanalyzed data on 14 participants from a study that had previously identified regions involved in lying (thus satisfying the criterion for sensitivity). We assessed the efficacy of functionally determined brain regions based on the lie–truth contrast for $N-1$ participants to detect deception in the Nth individual. Results showed that no region could be used to correctly detect deception across all individuals. The best results were obtained for medial prefrontal cortex (mPFC), correctly identifying 71% of participants as lying with no false alarms. Lowering the threshold for a response increased hits and false alarms. The results suggest that although brain imaging is a more direct index of cognition than the traditional polygraph, it is subject to many of the same caveats and thus neuroimaging does not appear to reveal processes that are necessarily unique to deception.

Historian Ken Alder (2007) has described the polygraph and lie detection as an "American obsession," a pursuit steeped more in an American cultural fantasy than reliable evidence. Traditional polygraphy, using an array of physiological measures such as heart-rate, blood pressure, and electrodermal response, faced several fundamental problems: the sensitivity of the test, the effectiveness of the test in measuring what it purports to measure (specificity), the generalizability of the test across measurement contexts, and its robustness to countermeasures. In 2003, the Committee to Review the Scientific Evidence on the Polygraph concluded:

> Notwithstanding the limitations of the quality of the empirical research and the limited ability to generalize to real-world settings, we conclude that in populations of examinees such as those represented in the polygraph research literature, untrained in countermeasures, specific-incident polygraph tests can discriminate

Correspondence should be addressed to: George T. Monteleone, Center for Cognitive and Social Neuroscience, University of Chicago, 5848 S. University Avenue, Chicago, IL 60637, USA

We thank Alex Japour for his contributions to this research. Support for this research was provided by the Department of Army Research Grant #DAAD19-03-1-0042 and the MacArthur Law & Neuroscience Project.

www.psypress.com/socialneuroscience DOI:10.1080/17470910801903530

lying from truth telling at rates well above chance, though well below perfection. (Committee to Review the Scientific Evidence on the Polygraph, 2003, p. 4)

New areas of research have since taken on the challenge of detecting deception by moving to neurophysiological measures, including the physiological detection of deception using functional magnetic resonance imaging (fMRI). The notion is that by imaging the brain processes involved in deception, the problems of traditional polygraphy can be avoided.

Several fMRI studies of lie detection have been conducted using binary-choice paradigms and nomothetic (group-level) analyses to determine whether it was possible to discriminate lying and truthful responding. Spence et al. (2001) reported a study in which participants either lied or told the truth about acts they had performed during the course of that day. A cue presented along with the question signaled to the participant whether to lie or tell the truth. Blood-oxygen-level dependent (BOLD) contrasts between lie and truth conditions revealed significantly greater activation during lying in the bilateral ventral prefrontal cortex (VLPFC), dorsolateral prefrontal cortex (DLPFC), medial prefrontal cortex (mPFC), and the left inferior parietal cortex. Spence et al. (2001) interpreted the results in terms of learning and motor response inhibition. Using an in-scanner adaptation of the Guilty Knowledge Test (GKT) with playing cards as stimuli, Langleben et al. (2002) instructed participants to either lie or tell the truth about cards they had in their possession. Results revealed a greater activation in the DLPFC and in the medial frontal gyrus extending to the anterior cingulate cortex (ACC) during lying than truth-telling. Also using the GKT, Phan et al. (2005) found lying to be associated with greater activity in the mPFC, bilateral DLPFC and VLPFC, and bilateral superior temporal sulci (STS). Activation in the mPFC and DLPFC were activated differentially by lying and truth-telling in all three studies.

Brain imaging has also been performed using experimental paradigms to simulate more realistic lying situations. Lee et al. (2002) used a forced-choice memory paradigm in which participants feigned memory impairments on two memory tests (digit memory task and autobiographical memory task) in a simulated assessment of malingering. The participants were requested to devise their own mental strategies for the lie, thereby increasing the external validity of the manipulation of lying. The activation for feigned memory impairment versus accurate recall conditions was similar for the two memory tasks and included anterior frontal regions (BA 9 and 10), bilateral parietal and temporal cortex, and subcortical regions including the caudate. The prefrontal region result suggested a role of information manipulation and executive control, with anterior regions (BA 10) participating in simultaneous processing of primary and secondary goals, and the DLPFC region processing anticipation of performance and working memory. The authors suggested that the activation of the parietal region was part of a network involved in calculated responses during lying, the temporal regions as participating in the cognitive manipulation of visual information, and the subcortical regions as involved in self-monitoring and inhibition of learned rules. A study by Nuñez, Casey, Egner, Hare, and Hirsch (2005) investigated the neural correlates of false versus true responses to autobiographical or non-autobiographical information. The investigators reported a contrast effect for autobiographical items in the mPFC, DLPFC, VLPFC, ACC, BA 9/10, and caudate. At the chosen levels of significance, however, no differences were found between lying and truth for non-autobiographical items, suggesting differences in neural response for lying depending on the content being tested. In an experimental study to assess the effects of the type of lie told, Ganis, Kosslyn, Stose, Thompson, and Yurgelun-Todd (2003) investigated lies that were rehearsed and spontaneous lies. Lies generally were associated with greater activation in BA 10, fusiform gyrus, and visual cortex whereas activation for lies was greater than truth-telling in mPFC and DLPFC only for spontaneous lies.

Across these studies, activity in the mPFC and DLPFC was greater for lying than truthful responding in all except the conditions in Ganis et al. (2003) in which the lies were autobiographical and well-rehearsed. It is not clear whether these regions are indicative of lying or are indicative of the increased self-monitoring and load on working memory that are often involved when one is lying. These studies have also focused on nomothetic analyses of the BOLD response as a function of lying versus truth-telling, where lying and truth-telling are not only known but are controlled by the experimenter. The critical question, which has limited the scientific validity of the traditional polygraph, is whether lying can be discriminated from truth-telling on an individual level based on the observed physiological response when one is answering the critical question.

The first fMRI study endeavoring to discriminate lying from truth-telling at the idiographic level was reported by Kozel et al. (2005). BOLD responses were used to develop a mathematical model weighting regions of interest (ROIs) to distinguish lying from truth-telling, and a second group's BOLD responses to the task were used to classify them as lying or truth-telling. The authors reported classification at 90% accuracy using a combination of analysis in the middle frontal gyrus, inferior frontal gyrus, and anterior cingulate. This is only slightly better than the classification accuracy found for traditional polygraphy. No single region showed classification at better than 85% accuracy, and each cluster had a false alarm rate of nearly 20%. The combined analysis yielded false alarms for the 10% who were not successfully classified.

In a similar analysis based on a set of group predictor clusters, Langleben et al. (2005) reported single-cluster predictions at no better than 75% correct classification using the area under the curve of the receiver operator characteristic function, which plots probability of a false positive (specificity) against probability of correctly predicting a lie (sensitivity.) The area under the curve was described as the probability of accurately classifying a pair of observations, with the best single region (left inferior parietal lobule) showing an area under the curve of 75.1%. Using a stepwise logistic regression model across 14 ROIs, the authors reported a final model area under the curve of 84.7%.

These initial attempts at individual classification of lies using fMRI were comparable to the results of the traditional polygraph: well above chance, but well below perfection (Committee to Review the Scientific Evidence on the Polygraph, 2003). Indeed, the value of the area under the ROC curve reported by Langleben et al. (2005) lies in the middle of the accuracy results from traditional polygraphy summarized in the report of the Committee to Review the Scientific Evidence on the Polygraph (2003).

More recently, Spence, Kaylor-Hughes, Brook, Lankappa, and Wilkinson (in press) scanned a woman who had been convicted of poisoning a child but who continued to profess her innocence. Results revealed longer response latencies and greater activation of the VLPFC and ACC when she endorsed her accuser's version of events than when she endorsed her version of events; the authors concluded that: "While we have not 'proven' that this subject is innocent, we demonstrate that her behavioural and functional anatomical parameters behave as if she were." Such a conclusion based on an individual's responses in an fMRI study can be questioned on logical grounds because lying is not the only cognitive process that has been associated with increased activation of the VLPFC and ACC, and little is known about how general the association is between lying and activation in these regions.

The purpose of the current study is to examine data from a typical fMRI study of lying (Phan et al., 2005) to determine how well lying could be differentiated from truth-telling at the idiographic level, with an emphasis on sensitivity, specificity, and generalizability across individuals. For instance, because the decision threshold can affect the sensitivity of measurement and the classification results, we used both a liberal and a conservative threshold for identifying significant differences in the BOLD response. In addition, prior studies examining idiographic classifications have restricted their analyses to the predictor clusters that significantly differentiated lying from truth-telling in the prediction sample. Such a procedure does not discriminate between the individual whose scan shows only that the ROIs identified in the predictor clusters were more active during lying than truth-telling—as set forth by the predictor model—and the individual whose fMRI scan shows that many if not most areas of the brain are more active during lying than truth-telling. Whether the inclusion of pseudo-ROIs increases or decreases specificity has not been examined previously. Consequently, we also performed analyses to determine the effects of the application of an ROI control region.

METHODS

A detailed account of experimental methods can be found in Phan et al (2005). Fourteen participants were given a modified version of the GKT in which they were asked to either tell the truth or lie about the possession of a playing card (e.g., 2 of hearts or 5 of clubs). Data were collected on a 4T Siemens platform, and preprocessed using Statistical Parametric Mapping (SPM), including slice-time correction, realignment, spatial normalization to the Montreal Neurological Institute (MNI) template, resampling of functional images to 2-mm isotropic voxels, and spatial smoothing with a 6 mm full-width–half maximum Gaussian kernel. Preprocessed data were converted from ANALYZE (spm99) to a format for use with the analysis software AFNI (Cox, 1996). A canonical hemodynamic response function was convolved with the experimental conditions using the AFNI tool WAVER, and this model was regressed against the experimental data at each voxel to provide within-subjects statistical maps of the responses for each condition, as well as a map of the lie vs. truth contrast.

Nomothetic analyses

A group response map was generated by carrying out a whole-brain voxelwise one-sample *t*-test of the Lie > Truth contrast value across all 14 participants at $p < .01$ with a cluster size extent threshold of 119 contiguous voxels based on corner-to-corner connectivity. Phan et al. (2005) reported results only in regions where they had *a priori* hypotheses, namely ACC, mPFC, DLPFC, and VLPFC. In this case, a whole-brain voxelwise result was generated to identify all possible regions that differentiated lies from true statements.

Idiographic analyses

Idiographic analyses were carried out at two significance levels to assess the accurate detection of lies on an individual by individual basis using a conservative ($p < .01$) and a liberal ($p < .05$) criterion for Type I errors. To develop the predictive model for the idiographic analysis, the contrast images for 13 of the 14 participants were entered into the second stage of a random-effects analysis (one-sample *t*-test, two-tailed; $t = 3.057$, df = 12, $p < .01$; $t = 2.179$, df = 12, $p < .05$). This process was performed for each of the 14 participants, resulting in 14 group response maps, each of which was then used as the template for the neural signature of lying when examining each individual contrast map. That is, after group analysis from 13 participants yielded predictor clusters, the remaining 14th participant's data was thresholded at the same individual voxel extent levels as the group data, yielding a set of target clusters for the individual. Specifically, each remaining individual's contrast map was submitted to the same threshold based on the coefficient of the least-squares estimate of the empirical data to the model ($p < .01$, $t = 2.585$ and $p < .05$, $t = 1.964$). Significant regions were determined by applying an individual voxel probability threshold of $p < .01$ with a cluster volume of 952 μl (119 contiguous voxels) and again with a voxelwise threshold of $p < .05$ with a cluster volume of 5160 μl (645 voxels) based on corner-to-corner connectivity in 3D space, which was equivalent to a connectivity radius of 3.46 mm. The cluster volume was chosen as the means to correct for multiple comparisons ($\alpha < .05$). Cluster volume threshold was determined with a Monte Carlo simulation for which the input parameters modeled the analysis (voxel size $2 \times 2 \times 2$ mm, connectivity radius 3.46 mm, individual voxel $p = .01$ or $p < .05$, Gaussian FWHM filter width = 6 mm) executed within a mask of the entire brain (231,766 voxels) for 5000 iterations using the AFNI program AlphaSim. The Monte Carlo simulation randomly simulates "active" voxels within the mask according the probability and spatial parameters for the specified number of iterations, ultimately calculating the probability that a cluster of size X would occur by chance. The volume X is then used as a selection criterion on the experimental data to obtain activity clusters that meet the corrected α level (Forman et al., 1995; Xiong, Gao, Lancaster, and Fox, 1995).

For each participant, individual target cluster maps were overlaid on group predictor maps to assess signal detection. Significant voxels from the individual maps that fell within the nomothetic predictor map were tallied as hits if the valences from individual results matched those of the group result. No predictor maps yielded significant True > Lie clusters; however, some target maps yielded True > Lie clusters. If a True > Lie target cluster overlapped with a Lie > True predictor cluster, it was counted as a false alarm. If no individual results were found for a predictor cluster, it was tallied as a miss for that cluster. If hit and false-alarm target voxels were detected in the same predictor cluster, the cluster was classified according to the result that had the larger volume.

A threshold was set for the size of clusters that overlapped between the nomothetic predictor map and individual signal map for each subject. Overlap cluster size was determined by executing a Monte Carlo simulation within the mask of the predictor map voxels at the appropriate voxelwise threshold ($p < .01$ or $p < .05$) to determine the cluster size corresponding to $\alpha < .05$. The cluster size for overlap regions at the voxelwise $p < .01$ level was 37 contiguous voxels, and at $p < .05$ it was 203 voxels.

The pseudo-ROI specificity control

Finally, we sought to discriminate between the individuals whose fMRI scan shows *only* that the predictor clusters were more active during lying than truth-telling and those who showed greater activation during lying than truth-telling across many areas, among them one or more of those specified in the predictor clusters. To do so, we formed a set of pseudo-ROIs that served as control regions for comparison with the predictor clusters. Assuming that the predictor clusters are in fact part of a reliable neural circuit, hit rates inside predictor clusters were hypothesized to be significantly higher than hit rates in the randomly sampled pseudo-ROIs. Given that results at the liberal threshold of $p < .05$ were expected to contain a considerably higher rate of false alarms than at $p < .01$, the $p < .05$ results were examined to determine whether the rate of false alarms would be decreased.

To form the pseudo-ROIs, the predictor ROIs were subtracted from a Talairach atlas map of all cortical grey matter, and for each predictor ROI a pseudo-ROI of equivalent volume was constructed from randomly sampled voxels in the remaining regions. Pseudo-ROIs were tested for signal detection in the same manner as group predictor ROIs, with a tally of significant voxels found at the individual level. A chi-square test (positive/negative × pseudo-ROI/actual-ROI) was used to compare frequency distributions of hits and false alarms in the pseudo-ROIs and predictor ROIs. Actual ROIs were expected to have a significantly higher hit rate than pseudo-ROIs, and predictor clusters that did not meet this criterion at $p < .05$ were discarded on grounds of non-specificity.

RESULTS

Nomothetic analyses

Whole-brain voxelwise group analysis showed that group Lie > Truth contrasts revealed significant activation in the medial prefrontal cortex (mPFC) in the region of the superior medial frontal gyrus, L inferior frontal gyrus (VLPFC), anterior cingulate, bilateral Brodmann's areas 9 and 10 (frontopolar region), bilateral temporo-parietal junction (TPJ), bilateral middle frontal gyrus (DLPFC), bilateral middle temporal gyurs, bilateral caudate, bilateral lingual gyrus, and the medial cingulate gyrus at a voxelwise $p < .01$ with a cluster minimum of 952 μl (119 contiguous voxels). Not all subjects' group analyses yielded the same set of regions; however, the majority of these regions (mPFC, VLPFC, ACC, BA 9.10, TPJ, R middle temporal, and L caudate) were found in all 14 group template maps. Results from a representative participant are illustrated in Figure 1 and detailed in Table 1.

Idiographic analyses

Figure 2 depicts the number of participants who showed activation in each ROI identified using data from the remaining 13 participants at two levels of detection threshold. First, more lenient detection thresholds are generally associated with more hits, fewer misses, and more false alarms. This was found in the present study: at $p < .01$, 43% of clusters were hits, 57% were misses, and 0% were false alarms, whereas at the more lenient detection threshold of $p < .05$, 53% were hits, 43% misses, and 4% of detected signals were false alarms. As would be expected from signal detection theory, decreasing the decision threshold increased the hit rate at the expense of also increasing the false alarm rate. Finally, no contrast was found in which truthful responses produced greater activation than lying.

The best classification at $p < .01$, with a classification success of 71%, was found in the mPFC in the region of the medial frontal gyrus (see Figure 2, top panel). No other regions predicted more than 7 of 14 participants. Four out of five subjects were classified in L Middle Temporal gyrus, but this region achieved significance and, thus, served as an ROI, in only 36% of the predictor maps. No false positives were observed in any of these regions.

Lowering the detection threshold to $p < .05$ increased the number of template maps in which the areas achieved significance and the percentage of correct classifications in some regions. For instance, 78% of participants were classified successfully as lying in mPFC and BA 9/10, and 64% in VLPFC, ACC, and Left TPJ (see Figure 2, middle panel). However, these

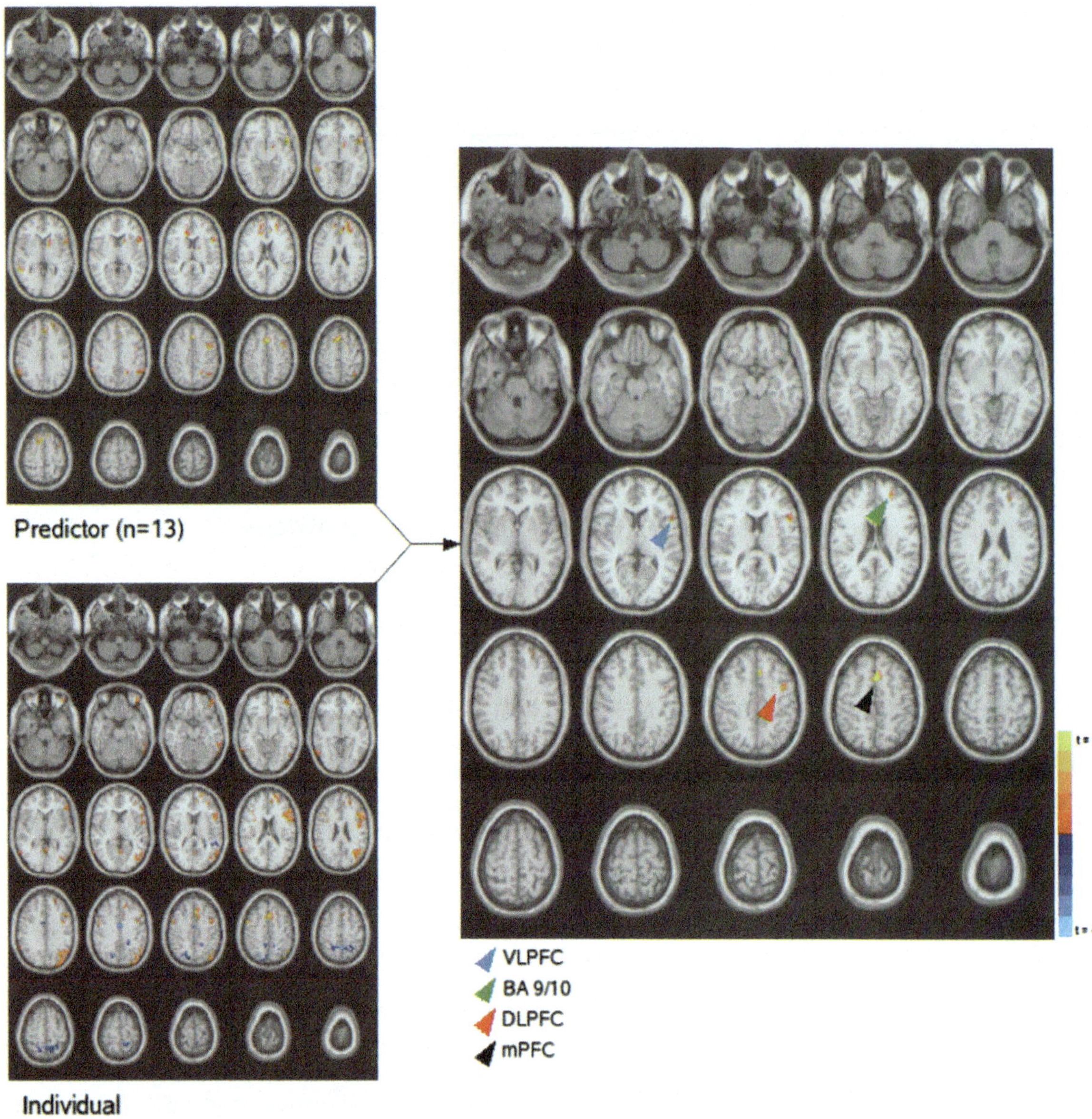

Figure 1. An illustration of the signal detection process from a representative subject. The Lie >True group predictor map (top left) was determined using 13 subjects' data, and was used to mask the individual Lie >True map (bottom left). Both maps were subjected to a voxelwise threshold of $p < .01$ with a minimum cluster size of 119 contiguous voxels as determined by a whole-brain Monte Carlo simulation at $\alpha < .05$. The overlap map was cluster-limited by another Monte Carlo simulation carried out in the predictor map, resulting in a cluster minimum of 37 voxels.

increased levels of detection came at the cost of increased false positives.

The effects of applying a pseudo-ROI control are illustrated in the bottom panel of Figure 2. If activation in a predicted ROI was accompanied by activation in the matched pseudo-ROI—that is, if the activation was nonspecific—then activation in the predicted ROI was treated as nondiagnostic. The results from four participants were affected by the implementation of pseudo-ROIs. The regions excluded were not entirely consistent across the participants, however. For participant #8, mPFC, BA 9/10, left TPJ, and left caudate regions were excluded, and for participant #14, right lingual gyrus was excluded. All five exclusions would otherwise have been considered hits.

DISCUSSION

Recent reports tout fMRI lie-detection as a means to spot liars (Peplow, 2004), detect terrorists (Wild,

TABLE 1

Lie > Truth regions for the representative subject pictured in Figure 1: The top table shows the predictor ROIs; the bottom table shows detected overlap regions

	Volume (μl)	*Hemisphere*	*Region(s)*	*TT atlas region(s)*	*BA*	*X*	*Y*	*Z*	*Mean B-value*
1	9104	R/L R/L L	mPFC BA 9,10 Anterior cingulate	Superior frontal gyrus Medial frontal gyrus Anterior cingulate	6, 9, 10, 32	4	47	26	.904
2	3544	L	VLPFC	Inferior frontal gyrus	47	−53	37	−18	.816
3	2552	L	Temporoparietal	Inferior parietal lobule	13, 40	−42	−36	26	.531
4	2120	R R	BA 9/10 Anterior cingulate	Medial frontal gyrus Anterior cingulate	10	14	61	2	.503
5	2040	L	DLPFC	Middle frontal gyrus	6	−42	23	31	.654
6	1664	L	Caudate	Caudate	25, 47	−15	29	−14	.548
7	1520	R	Temporoparietal	Superior temporal gyrus	42	60	−33	17	.660
8	1296	R	Middle temporal	Middle temporal gyrus	20	53	−25	−18	.855
1	1120	R/L	mPFC	Medial frontal gyrus	32	2	37	32	—
2	728	L	VLPFC	Inferior frontal gyrus	47	−53	37	4	—
3	672	L	BA 9/10	Superior frontal gyrus	9, 10	−24	68	7	—
4	480	L	DLPFC	Middle frontal gyrus	6	−22	27	38	—

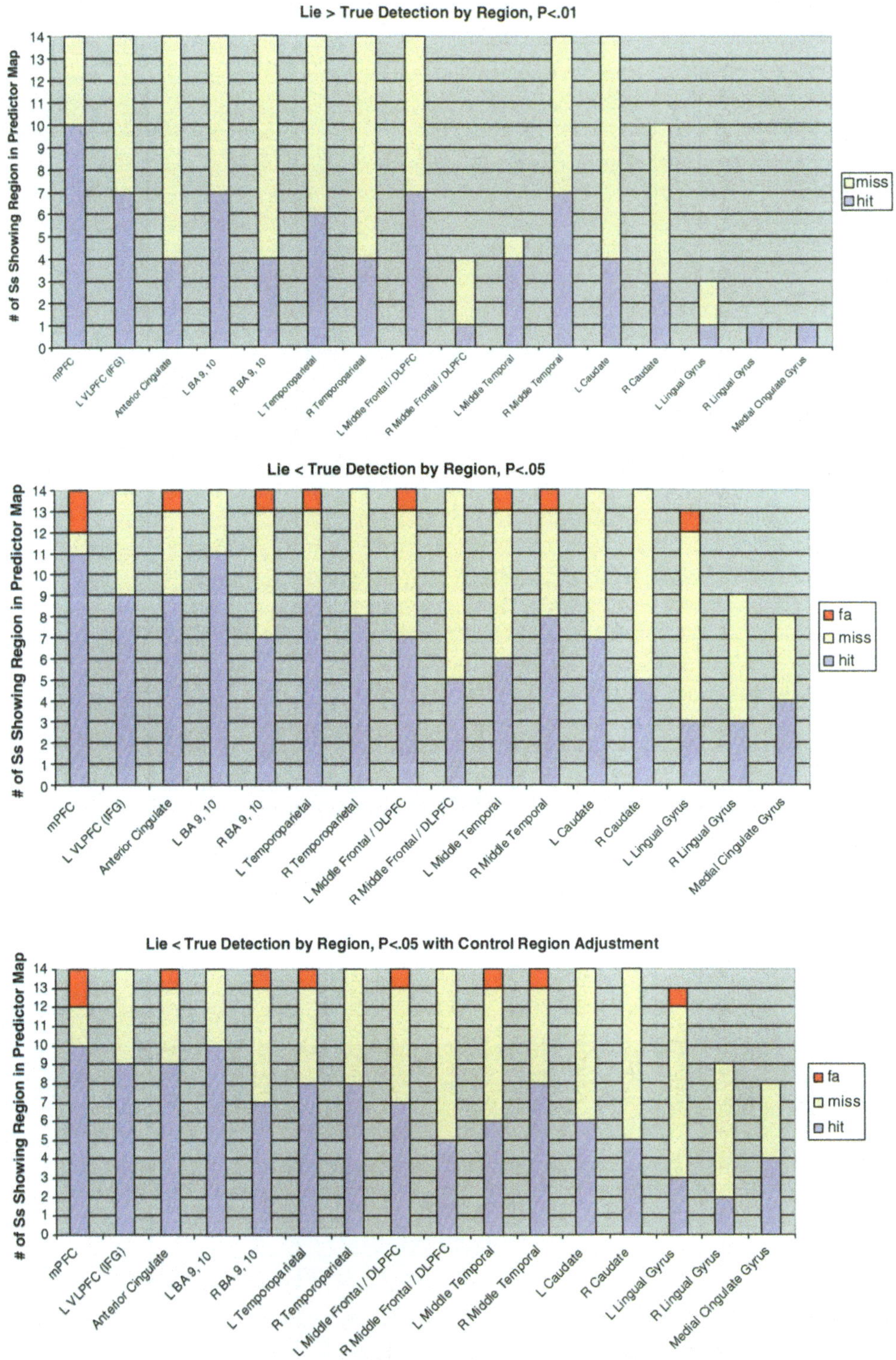

Figure 2. (Top panel) The number of hits and misses in each ROI at a significance threshold of $p < .01$ with a minimum overlap cluster size of 37 voxels. The total bar height indicates the number of subjects for which a response was predicted in each region. Some regions (R DLPFC, L MTG, R caudate, L and R lingual gyri, and medial cingulate) were predicted for only a fraction of subjects. No false alarms (True >Lie in a region predicting Lie >True) were detected at this significance level. Some regions (R DLPFC, L middle temporal, R caudate, L and R lingual, and cingulate gyrus) were significantly active in only a portion of subjects' predictor maps. (Middle panel) The number of hits and misses in each ROI at a significance threshold of $p < .05$ with a minimum overlap cluster size of 203 voxels. (Bottom panel) The number of hits, misses, and false alarms in each ROI at a significance threshold of $p < .05$ after the application of the pseudo-ROI control region (see text for an explanation of pseudo-ROIs).

2005), and reveal secret intentions in the brain (Haynes et al., 2007). Efforts to privatize new deception detection technologies have brought forth ethical concerns (Pearson, 2006), issues of personal privacy (Olson, 2005) and legal concerns about fourth amendment rights and personal liberty (Boire, 2005; Greely & Illes, 2007). Forensic applications and reports of effective "brain fingerprinting" (Farwell & Smith, 2001) and fMRI (Spence et al., in press) have been accompanied by efforts on the part of private interests to apply brain imaging for commercial gain. The present results raise several questions about the claim that there is an invariant neural locus unique to lying that is sensitive, specific, and consistent across individuals.

First, using a variant of the GKT, lying and truth-telling in the current study were associated with differential activation in as many as 16 different brain regions in some template maps. The classification of an individual as lying or not based on ROIs specified by the template maps proved to be well above chance, but no better than the levels typically observed using traditional polygraphy (Committee to Review the Scientific Evidence on the Polygraph, 2003). The best classification accuracy found for any ROI using a stringent classification threshold was 71% (with no false positives), found using the mPFC. Using a less stringent criterion for identifying areas of brain activation led to more hits, with activation in two ROIs in the frontal cortex producing correct classifications in 78% of cases, but false-alarm rates also increased across the areas that were identified.

Even though activation of the mPFC provided better than chance classification of individuals as lying in the present study, it does not mean that we detected a lie response *per se*. The mPFC has been associated with various cognitive processes including self-awareness and self-referent processing, mentalizing, executive functioning, and theory of mind (e.g., Frith & Frith, 2003; Krendl & Heatherton, in press; Saxe, Carey, & Kanwisher, 2004). Accurate classifications in the current study could have resulted from one or more of these cognitive processes occurring more often in the lying than in the truthful response conditions. It is obvious that knowing that lying leads to a skin conductance response does not logically mean that a skin conductance response indicates a person is lying, because lying is not the only process that produces a skin conductance response. Similarly, knowledge that lying leads to regional brain activation (e.g., mPFC) does not imply logically that activation of that region marks a person who is lying.

Brain imaging is a more direct index of cognition than the traditional polygraph, but it may be subject to many of the same caveats. For instance, we found that adopting a more lenient threshold increased hits, but at the expense of increasing false positives, just as has been found in studies of traditional polygraphy. False positives can be particularly pernicious in legal and employment applications, and even low false positive rates can lead to large numbers of truthful individuals being classified as lying when the base rate for lying is low. For instance, if the fMRI detection procedure has an accuracy index of .90 in a hypothetical population of 10,000 examinees that includes 10 liars, a detection threshold set to detect 80% of the liars will miss 2 of the 10 liars and will falsely classify 1,598 truth-telling individuals as lying (Committee to Review the Scientific Evidence on the Polygraph, 2003).

Second, our experimental paradigm made it possible to know when participants were telling the truth or lying, thereby making it possible to determine regions of brain activation that differentiated lying and truth-telling (i.e., a template). Typically, it is not possible to know whether a participant is telling the truth or lying. The intersubject variability in brain response to lying observed in the current study means that the template identified using a nomothetic approach may fail to characterize accurately many individuals' brain response to lying. This was evident in the present study, where nomothetic maps permitted lies to be accurately classified in the case of most but not all individuals. Templates designed at the idiographic rather than the nomothetic level may be worth investigating to discriminate lying from truth-telling, but the merit of such an approach has yet to be proven. To develop such a template, for instance, individuals would need to lie and to tell the truth in known sequences in response to multiple questions. This procedure may be limited if participants are not cooperative or if different kinds of lies are associated with different patterns of brain activation.

Third, the empirical support to date has raised questions about the notion that there is an invariant neural signature for lying. The card version of the GKT is a simple task that may limit individual variability in type of lie that is emitted. Although a card-based guilty knowledge test serves as a model for lying, it may be an overly simplified paradigm for testing legitimate lie-detection, and does not address the issue of differences in response according to different types of lie or the type of information involved. Thus, any level of success of lie classification based on this paradigm is but an early step towards more ecologically valid scenarios. Indeed, the research by Ganis et al. (2003) suggests that different types of lies may be associated with completely different patterns of brain activation. This seems plausible, as different types of lies are the product of one or more distinguishable component processes, which should be reflected in differences in neural activation.

Fourth, the brain imaging research on lying to date has not considered what detection problems might be introduced by participants who use physical or mental countermeasures. If a participant were to perform serial subtraction or to focus on suppressing an irrelevant response when responding truthfully and lying, for instance, would the pattern of neural activation that differentiates lying from truth-telling still be evident? Or, in light of the nomothetic template found in the present study, would thinking about the mental state of the examiner and oneself as the object of interrogation when responding truthfully act as a cognitive countermeasure to the detection of deception? Such questions will be important to address before the applied value of lie detection using fMRI can be determined.

Finally, social, cultural, and linguistic differences across participants in the understanding of questions and the meaning and appropriateness of deception may contribute to differences in regional activation patterns observed during lying versus truth-telling. In the absence of an invariant neural signature for lying, successful detection of deception using fMRI may require a better appreciation of possible sociocultural effects.

In sum, fMRI as a method for the detection of deception should carry the same burden of proof as demanded of any other method. As in prior studies, fMRI analyses permitted the differentiation of deceptive and truthful responding at the group level. The classification of lying based on contrast maps, however, was better than chance but far from perfect. Two different reasons for misclassifications were identified. In some cases, the activation of a predicted ROI did not reflect specific activation in this region but instead reflected activation of a very large area that included part of the predicted ROI. The use of pseudo-ROIs proved useful in identifying such cases. The second and more common reason for misclassifications was that one or more of the regions found to differentiate lying from truthful responding in the template map were not associated with significant differences in activation in the test participant. Using more lenient thresholds for detecting activation only nominally improved classification accuracy and came at the expense of increased false alarm rates. Closer inspection of these individuals' contrast maps (lie – true) suggested that a few subjects showed lying associated with very different regions of activation than most individuals. Although this is speculative, lying for these individuals may have been achieved by equally atypical means. Together, these results suggest that, although fMRI may permit investigation of the neural correlates of lying, at the moment it does not appear to provide a very accurate marker of lying that can be generalized across individuals or even perhaps across types of lies by the same individuals.

REFERENCES

Alder, K. (2007). *The lie detectors: The history of an American obsession*. New York: The Free Press.

Blasi, G., Goldberg, T., Weickert, T., Das, S., Kohn, P., Zoltick, B., Bertolino, A., Callicott, J., Weinberger, D. R., & Mattay, V. S. (2006). Brain regions underlying response inhibition and interference monitoring and suppression. *European Journal of Neuroscience, 23*(6), 1658–1664.

Blumenfeld, R. S., & Ranganath, C. (2006). Dorsolateral prefrontal cortex promotes long-term memory formation through its role in working memory organization. *Journal of Neuroscience, 26*(3), 916–925.

Boire, R. G. (2005). Searching the brain: the fourth amendment implications of brain-based deception detection devices. *American Journal of Bioethics, 5*(2), 62–63.

Committee to Review the Scientific Evidence on the Polygraph (2003). *The polygraph and lie detection.* Washington, DC: National Academy Press.

Cox, R. W. (1996). AFNI: Software for analysis and visualization of functional magnetic resonance neuroimages. *Computers and Biomedical Research, 29*, 162–173.

Farwell, L. A., & Smith, S. S. (2001). Using brain MERMER testing to detect knowledge despite efforts to conceal. *Journal of Forensic Sciences, 46*(1), 135–143.

Forman, S. D, Cohen, J. D., Fitzgerald, M., Eddy, W. F., Mintun, M. A., & Noll, D. C. (1995). Improved assessment of significant activation in functional magnetic resonance imaging (fMRI): Use of a cluster-size threshold. *Magnetic Resonance Medicine, 33*, 636–647.

Frith, D. D., & Frith, U. (2003). Interacting minds: A biological basis. *Science, 286*, 1692–1695.

Ganis, G., Kosslyn, S. M., Stose, S., Thompson, W. L., & Yurgelun-Todd, D. A. (2003). Neural correlates of different types of deception: an fMRI investigation. *Cerebral Cortex, 13*, 830–836.

Greely, H. T., & Illes, J. (2007). Neuroscience-based lie detection: the urgent need for regulation. *American Journal of Law & Medicine, 33*, 377–431.

Haynes, J. D., Sakai, K., Rees, G., Gilbert, S., Frith, C., & Passingham, D. (2007). Reading hidden intentions in the human brain. *Current Biology, 17*, 323–328.

Kozel, F. A., Johnson, K. A., Mu, Q., Grenesko, E. L., Laken, S. J., & George, M. S. (2005). Detecting deception using functional magnetic resonance imaging. *Biological Psychiatry, 58*(8), 605–613.

Kozel, F. A., Padgett, T. M., & George, M. S. (2004a). A replication study of the neural correlates of deception. *Behavioral Neuroscience, 118*, 852–856.

Kozel, F. A., Revell, L. J., Lorberbaum, J. P., Shastri, A., Elhai, J. D., Horner, M. D., Smith, A., Nahas, Z., Bohning, D. E., & George, M. S. (2004b). A pilot study of functional magnetic resonance imaging brain correlates of deception in healthy young men. *Journal of Neuropsychiatry and Clinical Neurosciences, 16*, 295–305.

Krendl, A. C., & Heatherton, T. F. (in press). Components of the social brain. In G. G. Berntson & J. T. Cacioppo (Eds.), *Handbook of neuroscience for the behavioral sciences.* New York: Wiley.

Langleben, D. D., Loughead, J. W., Bilker, W. B., Ruparel, K., Choldress, A. R., Busch, S. I., & Gur, R. C. (2005). Telling truth from lie in individual subjects with fast event-related fMRI. *Human Brain Mapping, 26*, 262–272.

Langleben, D. D., Schroeder, L., Maldjian, J. A., Gur, R. C., McDonald, S., Ragland, J. D., O'Brien, C. P., & Childress, A. R. (2002). Brain activity during simulated deception: An event-related functional magnetic resonance study. *NeuroImage, 15*, 727–732.

Lee, T. M., Liu, H. L., Tan, L. H., Chan, C. C. H., Mahankali, S., Feng, C. M., Hou, J., Fox, P. T., & Gao, J. H. (2002). Lie detection by functional magnetic resonance imaging. *Human Brain Mapping, 15*, 157–164.

Nuñez, J. M., Casey, B. J., Egner, T., Hare, T., & Hirsch, J. (2005). Intentional false responding shares neural substrates with response conflict and cognitive control. *NeuroImage, 25*, 267–277.

Olson, S. (2005). Brain scans raise privacy concerns. *Science, 307*, 1548–1550.

Pearson, H. (2006). Lure of lie detectors spooks ethicists. *Nature, 441*, 918–919.

Peplow, M. (2004). Brain imaging could spot liars. *Nature News.* Retrieved January 15, 2008 from http://www.bioedonline.org/news/news.cfm?art = 14 09

Phan, K. L., Magalhaes, A., Ziemlewicz, T. J., Fitzgerald, D. A., Green, C., & Smith, W. (2005). Neural correlates of telling lies: a functional magnetic resonance imaging study at 4 tesla. *Academic Radiology, 12*, 164–172.

Rahm, B., Opwis, K., Kaller, C. P., Spreer, J., Schwarzwald, R., Seifritz, E., Halsband, U., & Unterrainer, J.M. (2006). Tracking the subprocesses of decision-based action in the human frontal lobes. *NeuroImage, 30*(2), 656–667.

Saxe, R., Carey, S., & Kanwisher, N. (2004). Understanding other minds: linking developmental psychology and functional neuroimaging. *Annual Review of Psychology, 55*, 87–124.

Spence, S.A., Farrow, T.F., Herford, A.E., Wilkinson, I.D., Zheng, Y., & Woodruff, P.W. (2001). Behavioral and functional anatomical correlates of deception in humans. *NeuroReport, 12*, 2849–2853.

Spence, S. A., Kaylor-Hughes, C. J., Brook, M. L., Lankappa, S. T., & Wilkinson, I. D. (in press). 'Munchausen's syndrome by proxy' or a 'miscarriage of justice'? An initial application of functional neuroimaging to the question of guilt or innocence. *European Psychiatry.*

Wild, J. (2005). Brain imaging ready to detect terrorists, say neuroscientists. *Nature, 437*, 457.

Wirsing, B. (2007). *Revealing secret intentions in the brain (Press Release)*. Munich: Max Planck Society for the Advancement of Science.

Wolpe, P. R., Foster, K. R., & Langleben, D. D. (2005). Emerging neurotechnologies for lie detection: Promises and perils. *The American Journal of Bioethics, 5*(2), 39–49.

Xiong, J., Gao, J.-H., Lancaster, J. L., & Fox, P. (1995). Clustered pixels analysis for functional MRI activation studies of the human brain. *Human Brain Mapping, 3*, 287–301.

SOCIAL NEUROSCIENCE, 2009, 4 (6), 539–553

Psychology Press
Taylor & Francis Group

Neural processes underlying self- and other-related lies: An individual difference approach using fMRI

Giorgio Ganis

Harvard Medical School, Boston, Massachusetts, USA, and Massachusetts General Hospital, Charlestown, Massachusetts, USA, and Harvard University, Cambridge, Massachusetts, USA

Robert R. Morris

Massachusetts Institute of Technology, Cambridge, Massachusetts, USA

Stephen M. Kosslyn

Harvard University, Cambridge, Massachusetts, USA, and Massachusetts General Hospital, Boston, Massachusetts, USA

Two hypotheses were tested using a novel individual differences approach, which identifies rate-limiting brain regions, that is, brain regions in which variations in neural activity predict variations in behavioral performance. The first hypothesis is that the rate-limiting regions that support the production of lies about oneself (self-related) are partially distinct from those underlying the production of lies about other individuals (other-related). The second hypothesis is that a cingulate–insular–prefrontal network found to be rate-limiting for interference tasks is involved in both types of lies. The results confirmed both hypotheses and supported the utility of this individual differences approach in the study of deception in particular, as well in the study of complex cognitive phenomena more generally.

INTRODUCTION

Deception is a fundamental and pervasive social behavior that occurs when one individual attempts to convince another to accept as true what the deceiver believes to be false information in order to gain a benefit or avoid punishment (Vrij, 2000). Although in most cases deception has only minor negative results, or even positive ones (e.g., facilitation of social interactions), in some others it can have enormous adverse consequences (e.g., espionage).

Because of the potentially negative consequences of deception, many researchers have sought to understand deception and to devise methods to detect it. The traditional "emotional approach" to deception (Vrij, 2000) has led to methods based on observation of arousal-related changes in overt behavior (DePaulo et al., 2003) or on measurement of peripheral psychophysiological variables (e.g., skin conductance). However, such methods have shortcomings, from both a theoretical and an applied standpoint. From a theoretical standpoint, these methods typically

Correspondence should be addressed to: Giorgio Ganis, Martinos Center, Building 149, Massachusetts General Hospital, Harvard Medical School, Charlestown, MA 02129, USA. E-mail: ganis@nmr.mgh.harvard.edu

This material is based partially upon work supported by the National Science Foundation under Grant BCS-0322611. Any opinions, findings, and conclusions or recommendations expressed in this material are those of the authors and do not necessarily reflect the views of the National Science Foundation. The authors wish to thank Haline Schendan, Julian P. Keenan, and Mark D. Happel for helpful discussion, and Jonathan Weiss for help with stimulus design and implementation.

www.psypress.com/socialneuroscience DOI:10.1080/17470910801928271

overlook the fact that deception is not a single, unitary phenomenon; rather there are different types of lies (e.g., those made spontaneously vs. those based on a well-rehearsed and memorized scenario), and different mechanisms may underlie the different types (Ganis, Kosslyn, Stose, Thompson, & Yurgelun-Todd, 2003). From a practical standpoint, such methods are not very precise at detecting deception, in part because they lack specificity (National Research Council, 2003).

In an effort to circumvent these problems, work in the past three decades has emphasized the cognitive processes underlying deception (Buller & Burgoon, 1996; McCornack, 1992, 1997; Zuckerman, DePaulo, & Rosenthal, 1981), laying the foundation for the recent studies on the neurocognitive underpinning of deception. Early research on the neural bases of deception was conducted using event-related potentials (ERPs) (e.g., Allen & Iacono, 1997), but in the past few years there have been numerous functional magnetic resonance imaging (fMRI) studies of deception (Abe et al., 2006; Abe, Suzuki, Mori, Itoh, & Fujii, 2006; Davatzikos et al., 2005; Gamer, Bauermann, Stoeter, & Vossel, 2007; Ganis et al., 2003; Kozel et al., 2005; Kozel, Padgett, & George, 2004; Langleben et al., 2002, 2005; Lee et al., 2002, 2005; Mohamed et al., 2006; Nunez, Casey, Egner, Hare, & Hirsch, 2005; Phan et al., 2005; Spence et al., 2001, 2004).

With the advent of fMRI, researchers are beginning to characterize the mechanisms that underlie not only deception, but different sorts of deception. In a prior fMRI study, we put forward and tested the idea that, like most other complex cognitive processes, partially nonoverlapping neural systems support the generation of different types of lies (Ganis et al., 2003). In that study we documented differences between episodic autobiographical spontaneous lies versus rehearsed lies, and suggested that an additional distinction worth studying is between lies about oneself (self-related) and lies about other individuals (other-related). This distinction is important for at least two reasons. First, there are long-term differences between self-related and other-related lies because the vast majority of lies people tell appear to be self-related (DePaulo & Kashy, 1998); thus, most people are much more experienced in telling self- than other-related lies. Second, there is increasing evidence that the brain networks storing memories about oneself are partially distinct from those storing other types of memories, although the details are still unclear (e.g., Maguire, 2001; Schaefer et al., 2006; Svoboda, McKinnon, & Levine, 2006).

Thus, the first goal of this study was to use the same logic employed in Ganis et al. (2003) to compare the neural processes underlying the production of self-related lies and other-related lies. It is important to note that self-related lies are based on self-related memories, which are the same as autobiographical, but not as episodic, memories (see Roediger & Marsh, 2003): That is, although self-related memories often are about specific episodes (e.g., "I was at home on Sunday afternoon"), in many cases they are also about semantic or procedural information about oneself that is not associated with a specific episode (e.g., "I don't like classical music" or "I know how to ride a unicycle"). What is in common among all types of self-related memories is that they are about the self. Similarly, what is in common among all types of other-related memories is that they are about another person. Like self-related lies, other-related lies may be about episodic memories (e.g., a specific event during a televised Presidential debate) or about semantic memories not associated with specific episodes (e.g., "George Bush is a Republican").

Thus, the distinction between self-related and other-related lies is different from the classic distinction between episodic and semantic memory, and the neural mechanisms that underlie the two sorts of lies are not likely to be the same as those that underlie the two types of memories (cf. Tulving, 2002). Virtually all neuroimaging studies of deception have investigated lies about autobiographical events (either about a person's past life or about laboratory episodes). The only fMRI study that has examined nonautobiographical lies (in this case, not lies about other individuals, but about world knowledge—such as whether New York City is in Ohio) found no differences when comparing directly deceptive and honest responses (Nunez et al., 2005). In the same study, differences were found instead when comparing directly deceptive and honest responses for autobiographical lies (although no direct contrast was reported between the autobiographical and nonautobiographical conditions), including activation differences in the anterior cingulate cortex, and numerous prefrontal cortical foci. This suggests that distinct neural processes underlie these two types of lies.

To document differences between self- and other-related lies, in the current study we used a novel individual differences logic that employs a brain–behavior correlation method (Ganis,

Thompson, & Kosslyn, 2005; Kosslyn, Thompson, Kim, Rauch, & Alpert, 1996; Miller et al., 2002; Ng et al., 2001; Plomin & Kosslyn, 2001; Wager et al., 2005). The core of this logic revolves around the distinction between *rate-limiting* and *minimally sufficient processes.* Rate-limiting processes for a given task are performance bottlenecks; that is, task performance depends greatly on the efficiency of these processes. In contrast, performance is not greatly affected by the efficiency of minimally sufficient processes. An analogy can be used to clarify these concepts. The precision with which one can produce a miniature airplane model depends critically on one's fine motor skills – a rate-limiting process for this task – but not on the strength of one's biceps, assuming the minimal strength required to support the arm. In contrast, the maximum weight one can lift depends crucially on the strength of one's biceps, but only minimally on fine motor skills. This analogy serves to underscore that a rate-limiting process in one task may not be one in another; the key is the role that the process plays in enhancing performance of a specific task. Moreover, we must note that virtually all complex tasks (e.g., throwing darts) depend on combinations of minimally sufficient and rate-limiting processes.

One way to identify the neural substrates of rate-limiting processes for a certain task is to find brain regions where activation predicts task performance across individuals. Such brain regions, by definition, are the ones that mostly affect variations in performance. Although the resulting evidence is still correlational, using information from individual differences to identify rate-limiting processes brings us closer to the goal of defining the sets of component processes that give rise to performance in complex tasks such as deception. Indeed, Sidtis (2007) examined neuroimaging data of speech production, and found that regions revealed by individual differences in performance correspond better to the regions usually found to cause speech problems when disrupted by brain damage than do regions revealed by standard contrast analyses.

Most neuroimaging studies have focused on minimally sufficient processes by finding areas that are consistently more activated in one condition than another. In contrast, only a relatively small number of fMRI studies, none of them investigating the neural bases of deception, have focused on rate-limiting processes and on documenting differences in rate-limiting processes between conditions (e.g., Ganis et al., 2005; Kosslyn et al., 1996; Miller et al., 2002; Ng et al., 2001; Wager et al., 2005).

Among the processes thought to be recruited during the generation of deceptive responses are those required to monitor and resolve interference caused by a "prepotent" response—namely, the response that corresponds to the truth (e.g., Johnson, Barnhardt, & Zhu, 2004; Vrij, 2000; Zuckerman, DePaulo, & Rosenthal, 1981). This response must be suppressed in order to produce a lie. To date, there have been no neuroimaging studies that address whether these interference-monitoring and resolution processes are rate-limiting for deception. However, a study of interference-monitoring and resolution processes using the same individual difference logic used here found a network of rate-limiting regions comprising the cingulate, the insula, and portions of the prefrontal cortex (Wager et al., 2005): Activation in these regions predicted the response times (RTs, in this case difference scores between interference and no-interference conditions) in Go/No Go, flanker task, and stimulus–response compatibility tasks, all requiring some form of interference monitoring and resolution.

Thus, the second goal of this study was to test the hypothesis that interference-monitoring and resolution processes are rate-limiting for deception tasks. We expected to find the same network of rate-limiting brain regions found by Wager et al. (2005) when we conducted individual differences analyses of the data from our deception conditions.

MATERIALS AND METHODS

Participants

Sixteen Harvard University undergraduates (8 females, 8 males; mean age 23 years), volunteered to take part in the study for pay. All had normal or corrected-to-normal vision, no history of neurological disease, and were right-handed. All participants gave written informed consent for the study according to the protocols approved by Harvard University and Massachusetts General Hospital Institutional Review Boards. We analyzed data from 14 participants. Data from 2 participants (1 female and 1 male) were not analyzed because they did not complete the study; the demographics of these 2 participants were comparable to those of the entire group.

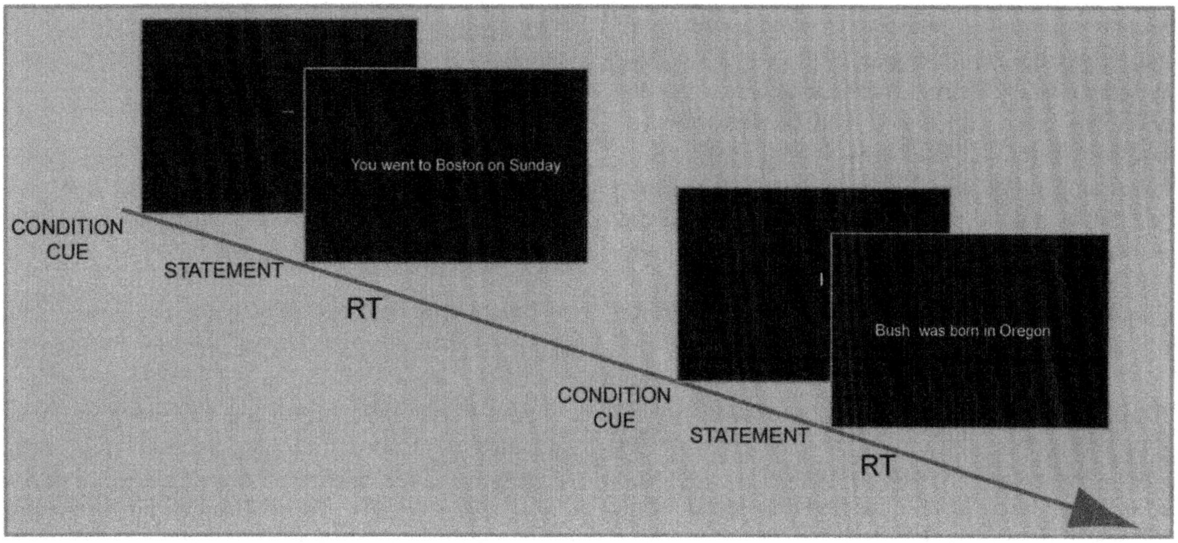

Figure 1. Schematic illustration of two experimental trials. The time between the condition cue and the statement was 4 s. The time between the response (and collection of response time, RT) and the next condition cue was 14 s.

Stimuli

The stimuli were 200 written statements, 100 about oneself (self-related) and 100 about George W. Bush (other-related). George W. Bush was chosen for the other-related condition because participants were highly familiar with numerous facts about him and his life from the media, but they had never met him in person. With others, such as close friends or relatives, most other-related knowledge would be contaminated by self-related memories of interactions with these people.[1]

Half of the statements were true and half false. Each true sentence had a companion false sentence (e.g., "You went to England last week" vs. "You went to France last week") so that the response keys were not systematically associated with honest or deceptive responses: Key 1 (or 2) was the appropriate response for half of the honest responses, and for half of the deceptive responses. For the self-related statements, a week before the fMRI session, the experimenter interviewed each participant about numerous topics so that true and false statements could be constructed (e.g., "Last year you went to England"). The specific content of these statements was different for each participant. The other-related sentences were well-known facts about George W. Bush (as assessed by polling a group of undergraduates who did not participate in this study), and were identical for all participants. During a second session, also preceding the fMRI session, participants were shown all the statements and asked to cross out the false ones, to ensure they knew the correct answers. Statements were kept simple and short and their average length (5.4 words) was matched across conditions.

Procedure

Tasks were administered on a Macintosh G3 Powerbook computer using Psyscope software (Macwhinney, Cohen, & Provost, 1997). Stimuli were projected via a magnetically shielded LCD video projector onto a translucent screen placed behind the head of the participants. A front-surface mirror mounted on the head coil enabled viewing of the screen. Prior to the MRI session, we administered general health history and Edinburgh Handedness (Oldfield, 1971) questionnaires.

At the outset of the fMRI session, participants read instructions on the computer screen and paraphrased them aloud. We corrected any misconceptions at this time. We then administered 10 practice trials, using simple statements not used in the study (e.g., "Bush is a singer"). Before each statement a condition cue was shown, either a small horizontal line or a small vertical line (Figure 1).

[1] Data collection for this study ended in the second half of 2005, prior to the widespread negative feelings towards President Bush that arose when the war in Iraq took a turn for the worse. Thus, at the time the study was conducted, the self/other distinction was not clearly confounded with major differences in emotional valence.

The cue instructed participants whether to lie or tell the truth on that trial. The cue type was counterbalanced across participants, so that the horizontal (vertical) line prompted deceptive responses in half of the participants and honest responses in the other half. Participants pressed one key to indicate that the statement was true, and a second key to indicate it was false. They were instructed to respond as quickly as possible without sacrificing accuracy. RTs were measured relative to the onset of each statement.

The fMRI session consisted of eight functional scans. During these scans we presented all 200 statements, in a random order. Each trial began with the condition cue, lasting 4 s. Next, a statement appeared and participants pressed one key to indicate that the statement was true, or a second key to indicate that it was false. After the key press, an intertrial fixation dot was shown for 14 s (to allow the hemodynamic response to return to baseline) before the presentation of the cue for the next trial (Figure 1).

To ensure that participants were actually trying to deceive another person, and gain something for themselves (consistent with our definition of deception), we devised a monetary game. Participants were told that during the fMRI session an investigator would be observing their eye movements and facial expressions by using a camera placed in the scanner, to determine whether they were lying on each trial. Participants would receive an additional 25 cents each time they lied successfully and would lose 25 cents each time their lie was caught by the investigator. All participants received the maximum amount at the end of the study, according to the IRB requirements.

MRI parameters

We used a 3T Siemens Allegra scanner with whole head coil. We collected full-volume structural images using an MPRAGE sequence, before the functional scans (128, 1.3 mm thick sagittal slices, 256×256 matrix). Blood oxygenation changes were assessed with functional scans using a T2*-sensitive sequence (gradient echo, TR = 2000 ms, TE = 30 ms, FOV = 20 cm, flip angle = 90°, 64×64 matrix, voxel size = $3.125 \times 3.125 \times 5$ mm). Each scan resulted in 270 volumes, each composed of twenty-one 5-mm oblique slices (slice gap = 1 mm).

Analyses

Behavioral data

We analyzed mean RTs and error rates (ERs) using a repeated-measures ANOVA with two factors: Deception (lie vs. truth) and Content (self-related vs. other-related).

FMRI data

Data were preprocessed and analyzed with AFNI (Cox, 1996) using the following sequence of steps: (1) slice timing correction; (2) motion correction; (3) spatial smoothing with a Gaussian filter (full-width half-maximum = 4 mm); (4) amplitude normalization, by scaling each time-series to a mean of 100 and calculating the percent change about this mean; (5) spatial normalization to the MNI305 template (Collins, Neelin, Peters, & Evans, 1994); and (6) spatial resampling to a $3 \times 3 \times 3$ mm grid. To model the hemodynamic response function for correct trials in each condition, we used a family of 10 tent functions (covering the period between 4 s before the statement onset to 16 s after it) and estimated the fMRI response at each time point using multiple linear regression. The multiple regression model included fourth-order polynomial trend regressors for each scan to model slow variations in blood-oxygen-level dependent (BOLD) signal over time. In addition, for each condition (self–honest, self–deceptive, other–honest, other–deceptive) there was one regressor for each of the 10 tent functions. Incorrect trials, regardless of the condition they came from, were modeled by using a single set of regressors. Maps of percent signal change for each participant and condition were obtained using the normalized regression coefficients between 6 and 9 s after the onset of each statement, encompassing the peak of the hemodynamic response.

The analyses were similar to those conducted in Ganis et al. (2005). First, we calculated the z-scores of the difference between the deceptive and honest conditions for the RTs, independently for the self- and other-related conditions. Second, the same z-scores were calculated for the fMRI data at each voxel. Difference scores were used because there are large overall individual differences in speed and accuracy, and these would mask any effects of deception *per se* if not contrasted with a within-individual baseline. The normalization procedure enabled us to compare the findings across regions by using a common scale. Third, we identified which brain regions predicted the behavioral z-scores across participants by conducting a correlation analysis. For this analysis, we used clusters with a 10-voxel extent threshold, all significant at $p < .005$ (corresponding to $|r| > .7$). These parameters provide a good compromise between sensitivity and protection against false positives (Xiong, Gao, Lancaster, & Fox, 1995). To ensure that the results were not due to outliers, for each of the resulting clusters we conducted the correlation analysis a second time using the robust regression algorithm provided by STATA's command "rreg", which employs an iteratively reweighted least-squares procedure (StataCorp, 2001). At most one data point for each condition (on about 5% of cases) was given a value of zero by the weighting procedure (i.e., it was flagged as a potential outlier). Finally, we entered the data from identified brain regions into a stepwise forward multiple regression, independently for each condition, to determine which regions were driving the correlations.

The specificity of the results was tested using three methods. First, we determined the extent to which the fMRI data predicted the behavioral data in one condition, by using the brain regions defined by the analysis in the other condition. That is, the brain regions found in the correlation analysis for the self-related condition were used to attempt to predict the data in the other-related condition (and vice versa). If the brain regions found in one condition are specific, activation in them should not predict the RTs in the other condition. Furthermore, the correlation coefficients in the two cases should be significantly different when compared directly with each other. Second, to ensure that the results were not due to global correlations between brain activation and behavior, we used the average brain activation (after removing the brain regions found in the correlation analyses) as a covariate in a partial regression analysis. Third, to test the temporal specificity of the results, we conducted the same correlation analyses in an earlier time window (0–4 s after condition cue presentation), before the neural sequelae elicited by deception processes were evident in the hemodynamic response.

RESULTS

Behavioral data

On average, participants responded correctly to 94.7% of the statements, and responded more quickly on correct trials (Figure 2) in the honest than in the deceptive condition (2073 ms and 2302 ms, respectively; $F(1, 13) = 16.4$, $p < .001$), consistent with the behavioral results reported by Morgan, Tolley, & Kosslyn (in press) with a very similar paradigm that measured voice onset. In addition, participants responded more quickly in the self- than other-related condition (2117 ms and 2258 ms, respectively; $F(1, 13) = 18.7$, $p < .001$). The size of the deception effect (honest minus deceptive) was numerically smaller for self- than other-related conditions (216 ms and 244 ms, respectively), but the difference was not significant, $F(1, 13) = 0.23$, $p > .5$.

In addition, the participants made fewer errors (Figure 3) in the honest than in the deceptive condition (3.7% and 6.9%, respectively; $F(1, 13) = 6.9$, $p < .05$), but they made comparable numbers of errors in the self- and other-related conditions (5.2% and 5.4%, respectively; $F(1, 13) = 0.67$, $p > .1$). Similarly to the RTs, the size of the deception effect in the ERs did not differ between self- and other-related conditions (3.2% and 3.3%, respectively; $F(1, 13) = .98$, $p > .1$). The largest correlation between RTs and ERs was in the self-related honest condition, but this correlation was not significant ($r = .37$, $p = .19$). All other correlations between RTs and ERs were close to

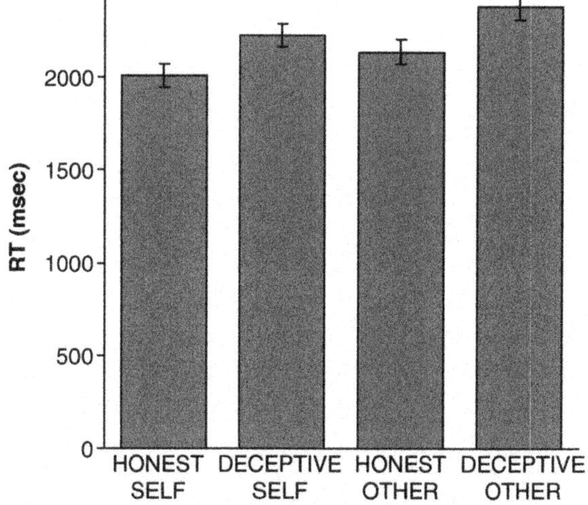

Figure 2. Average RTs in the four experimental conditions. Error bars indicate between-participant variability.

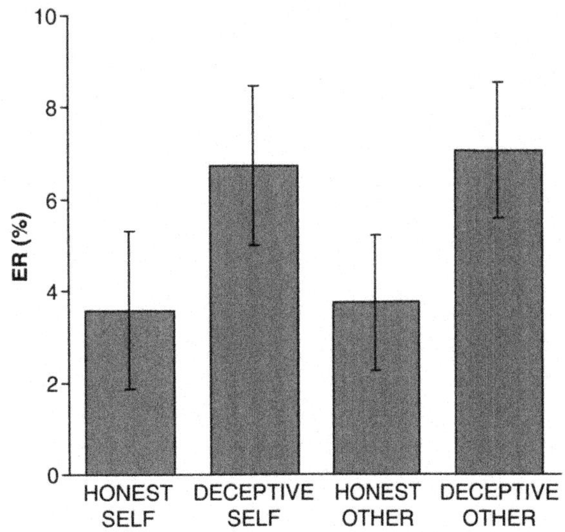

Figure 3. Average error rates (ERs) in the four experimental conditions. Error bars indicate between-participant variability.

zero (all r values $< .15$), indicating that the results were not due to speed–accuracy trade-offs. We did, however, find that participants who tended to have a large deception effect in the self-condition also tended to have a large effect in the other-condition, as indicated by a positive correlation over participants between the RT deception effect in the two conditions ($r = .58$, $p < .05$). The correlation between the ER deception effect in the two conditions was not significant ($r = .2$, $p > .4$).

FMRI data

Given the focus of this article, we only report the results of the analysis of rate-limiting processes. The results of the main correlation analyses are given in Tables 1 and 2 and shown in Figures 4 and 5. Brain activation predicted the RTs in numerous regions during the self-related condition. In many cases, activation in single clusters predicted over 65% of the variance in the RTs. In the frontal lobe, we found clusters in the cingulate gyrus (GC), the medial, inferior, middle, and superior frontal gyri (GFd, GFi, GFm, and GFs, respectively), and the insula (INS). In the occipital and temporal lobes, we found clusters encompassing portions of the lingual gyrus (GL), the inferior and middle occipital gyri (GOi and GOm, respectively), the fusiform gyrus (GF), and the parahippocampal gyrus (Gh) and hippocampus (Hi). In the parietal lobe, we found clusters in the precuneus (PCu) and the cuneus (Cun). Finally, we also found clusters in the right thalamus (Th) and in the left cerebellum (Cer). The correlation between brain activation and behavior for all clusters was positive (Table 1 and Figures 4 and 5); that is, high deception z-scores for brain activation were associated with high deception z-scores for the RTs.

Brain activation also predicted the RTs in numerous regions during the other-related condition (Table 2 and Figures 4 and 5). Similarly to the self-related case, activation in some clusters predicted over 65% of the variance in the RTs. In the frontal lobes, clusters included the GC, the INS and the claustrum (Cl), the GFm and the GFs. In the temporal lobe, we found a cluster in the left Gh and Hi. Finally, in the parietal lobe we found clusters in the left PCu and inferior parietal lobule (LPi). The GC, INS, and GFm/GFs clusters were very close to, but did not overlap with, those found in the self-related condition. The correlation between brain activation and behavior for all clusters was negative (Tables 1 and 2 and Figures 4 and 5); that is, larger deception z-scores for brain activation were associated with low deception z-scores for RTs.

To determine which brain regions were driving the pattern of correlations with the RTs, we entered the identified brain regions (hereafter, regions of interest, ROIs) into a stepwise forward multiple linear regression analysis, independently for the two conditions, as in prior work (Ganis et al., 2005; Kosslyn et al., 1996). However, due to collinearity (i.e., most variables were highly correlated with one another), it was not possible to obtain reliable estimates of the relative contribution of the different variables. This suggests that all regions may function together as a tightly coupled network: The average correlation among areas was .76 and .78 for the self-related and other-related conditions, respectively). To partially circumvent this problem, we used an alternative approach. We calculated multiple linear regression models for all possible subsets of three areas and determined which subset explained the largest amount of variance in the RTs. For the self-related lies, the results indicated that the subset comprising the GC/GFd/GFs, the GL/GOi/Cun/GF/GOm, and the Thal clusters was the most predictive (Table 1): Together, this subset of regions explained 94.9% of the variance in the RTs. For the other-related lies, the subset encompassing the GFm/GFs, the GC (BA 32/24), and the GFm clusters was the most predictive (Table 2): Together, this subset of regions explained 81% of the variance in the RTs.

TABLE 1

Brain areas that had significant correlations between BOLD activation and response times (self-related deception minus honest responses for both measures)

Regions within cluster	*Volume (mm^3)*			*Correlation*			*Talairach coordinates (center of mass)*			*Talairach coordinates (range)*		
Name	*BA*		*r*	*x*	*y*	*z*	*Min x*	*Max x*	*Min y*	*Max y*	*Min z*	*Max z*
GC/GFd/GFs	32/24/8	3024	.84	−3	17	38	−14	8	8	32	33	48
GC/GFd	32/24/9	675	.85	6	34	21	2	11	29	38	12	30
GC/GFd	32/9	486	.82	5	41	9	2	11	41	44	3	15
GFi/INS/Cl	47/13	945	.83	34	21	−3	26	44	17	26	−16	6
INS/Cl/GFi	13/47/45	918	.81	−30	19	4	−35	−26	14	26	−4	12
GFs/GFm	10/9	2511	.87	31	49	18	20	41	41	53	6	30
GFm/GFs	9/10	270	.78	−32	40	26	−35	−32	38	44	21	30
GL/GOi/Cun/GF/GOm	18/17	3618	.84	21	−96	−8	11	32	−104	−89	−22	12
GOi/GF/GL/Cun	18/17	2430	.90	−26	−93	−8	−35	−20	−101	−86	−16	3
GL/Cun	18/17	2430	.91	7	−88	−1	−2	14	−95	−80	−16	18
Cun/GL	18/17	2214	.88	9	−74	−5	−2	17	−83	−68	−22	9
Cun	19/18	1053	.75	−11	−86	27	−17	−8	−92	−80	18	36
GL	18	999	.88	−10	−73	−8	−14	−5	−77	−68	−19	3
PCu/Cun	31/18	378	.74	12	−71	24	8	14	−74	−68	21	27
Gh/Hi	36	205	.92	−29	−26	−10	−32	−26	−29	−23	−13	−7
Th	N/A	297	.80	3	−20	7	2	5	−23	−20	3	12
Cer	N/A	567	.87	−11	−78	−27	−17	−5	−80	−77	−31	−25
Cer	N/A	378	.77	−10	−62	−5	−17	−5	−65	−62	−7	−4
Cer	N/A	297	.75	−28	−75	−20	−32	−23	−77	−74	−22	−19

Notes: Abbreviations: BA, Brodmann's area; GC, cingulate gyrus; GFd, medial frontal gyrus; GFs, superior frontal gyrus; GFi, inferior frontal gyrus; INS, insula; Cl, claustrum; GOi, inferior occipital gyrus; GF, fusiform gyrus; GOm, middle occipital gyrus; GL, lingual gyrus; Cun, cuneus; PCu, precuneus; Gh, parahippocampal gyrus; Hi, hippocampus; GFm, middle frontal gyrus; Th, thalamus; Cer, cerebellum. Correlations were calculated with a robust regression routine from the software package STATA (StataCorp, 2001). Coordinates: *x*, left/right; *y*, anterior/posterior; *z*, inferior/superior. Shown in italics are the three regions that, together, explained variance in the RTs better than any other subset of three regions.

TABLE 2

Brain areas that had significant correlations between BOLD activation and response times (other-related deception minus honest responses for both measures)

Regions within cluster	Volume (mm^3)			*Correlation*			*Talairach coordinates (center of mass)*				*Talairach coordinates (range)*	
Name	*BA*		*r*	*x*	*y*	*z*	*Min x*	*Max x*	*Min y*	*Max y*	*Min z*	*Max z*
GC	24	324	−.80	5	−4	39	2	8	−5	−2	36	45
GC	32/24	297	−.87	13	9	39	11	17	8	11	36	45
INS/Cl	13	729	−.86	−35	8	9	−41	−32	5	14	6	18
INS/Cl	13	459	−.82	33	0	3	32	35	−2	2	−4	9
INS	13/41	378	−.80	−44	−23	18	−47	−41	−26	−20	15	21
GFm/GFs	10/46	351	−.70	−42	49	18	−47	−41	47	53	15	21
GFm	8/9	270	−.78	−33	31	38	−38	−32	29	35	36	42
PCu	7	432	−.86	17	−51	50	8	23	−53	−50	45	54
LPi	40	270	−.79	34	−37	42	32	38	−41	−35	39	45
Gh/Hi	36	513	−.78	−28	−12	−19	−32	−26	−17	−8	−22	−19

Notes: Abbreviations: BA, Brodmanns area; GC, cingulate gyrus; INS, insula; Cl, claustrum; GFm, middle frontal gyrus; GFs, superior frontal gyrus; PCu, precuneus; LPi, inferior parietal lobule; Hi, hippocampus; Gh, parahippocampal gyrus. Correlations were calculated with a robust regression routine from the software package STATA (StataCorp, 2001). Coordinates: *x*, left/right; *y*, anterior/posterior; *z*, inferior/superior. Shown in italics are the three regions that, together, explained variance in the RTs better than any other subset of three regions.

The effect of global brain activation on the results was assessed by measuring the partial correlation between the RTs and the brain activation in the ROIs, holding the global brain activation variable constant. The sign and significance of the partial correlation coefficients for all ROIs were the same as in the original analysis, indicating that global brain activation was not a factor.

Next, to test the specificity of the results, for each ROI we compared the correlation coefficient we found in one condition with that calculated by using the same ROI to predict the RTs in the other condition. If the ROIs in one condition were not specific then they would also predict the RTs in the other condition. The results were clear: No correlation coefficient was significant when trying to predict the RTs for one condition using the ROI from the other condition (all $|rs| < 0.3$). Furthermore, all correlations between the ROIs and the RTs in the corresponding condition were significantly larger than those with the RTs from the other condition ($p < .01$ in all cases).

To provide evidence for the temporal specificity of the correlations, we conducted the same analyses on an earlier time window of the data, between 0 and 4 s after the onset of the sentence. Given the typical hemodynamic delay (e.g., Kruggel & von Cramon, 1999), we did not expect any ROIs to predict behavior in this time window. The results confirmed this prediction: With the same threshold and cluster size used in the main analysis, activation in none of the brain regions predicted performance. This result also confirmed that the parameters used in these analyses were unlikely to produce false positives.

Several brain regions found in the self-related condition overlapped with those of the cingulate–insular–prefrontal network for interference monitoring and resolution found in the study by Wager et al. (2005). These included the main GC cluster ($x = 4$, $y = 11$, $z = 45$ in Wager et al.'s study), the two INS clusters ($x = -34$, $y = 19$, $z = 5$, and $x = 34$, $y = 19$, $z = 5$ in Wager et al.'s study), and the two GFm/GFs clusters ($x = -30$, $y = 49$, $z = 20$, and $x = 30$, $y = 41$, $z = 15$ in Wager et al.'s study). A cingulate–insular–prefrontal network was also found for the other-related condition, but the coordinates of the brain regions involved did not overlap exactly with those found in the self-related lie condition. These comparisons provide additional evidence that the two types

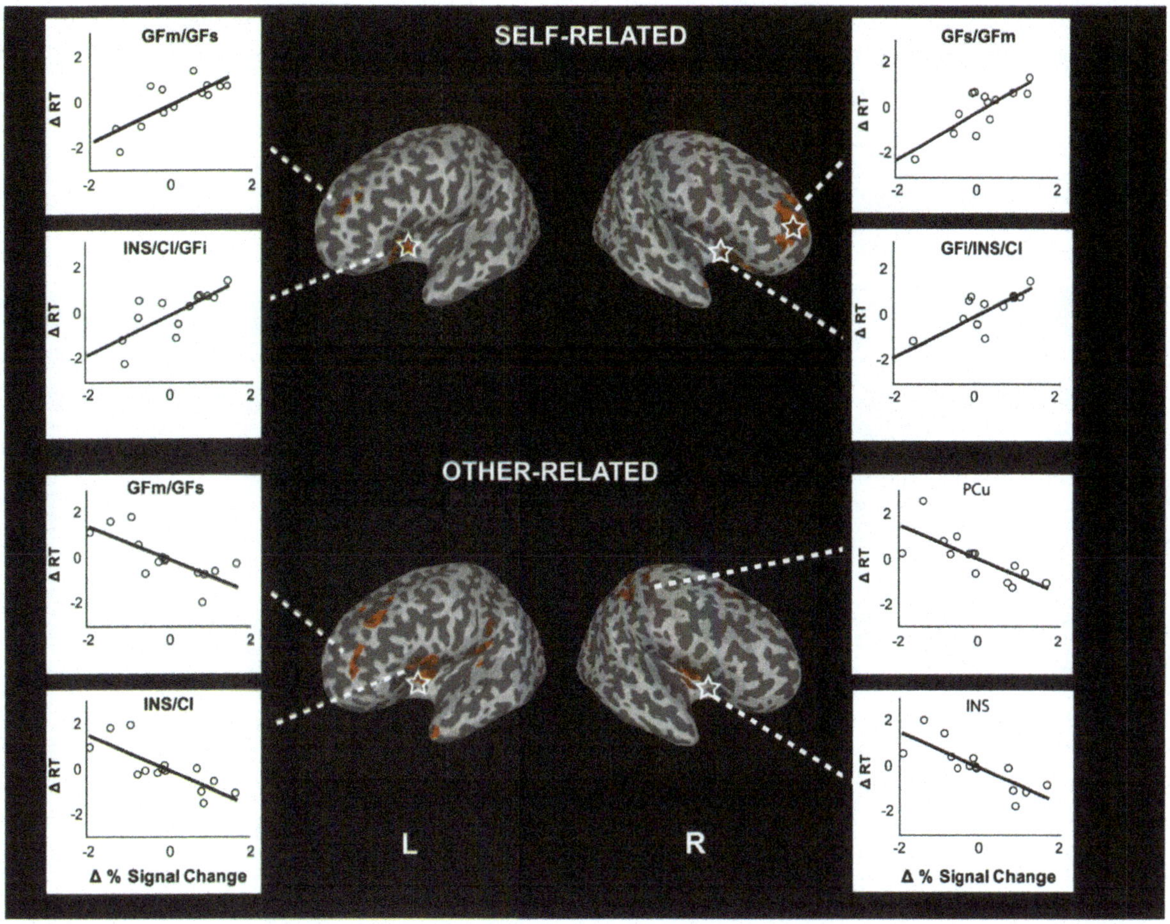

Figure 4. Lateral view of the results of the main correlation analysis identifying regions that predicted RTs across participants. Only clusters with more than 10 adjacent voxels at $p < .005$ (corresponding to $|r| > .7$) were considered significant (Tables 1 and 2). To illustrate clearly regions in which there was significant activation, the clusters shown here were drawn with a lower threshold ($p = 0.01$). Regions that predicted the RTs in the self-related (top) and other-related (bottom) conditions are shown in red. The stars on the insula and on the prefrontal cortex indicate the coordinates of the center of mass of the corresponding rate-limiting regions found by Wager et al. (2005) during three interference tasks. The scatterplots show RTs' z-scores as a function of brain activation z-scores for representative regions.

of lies, about the self vs. about another person, rely on distinct neural systems.

DISCUSSION

First and foremost, we found that the self-related and other-related conditions engaged different sets of rate-limiting regions. Although some subsets of these regions are similar, the two sets are clearly distinct. Crucially, these differences cannot be attributed to differences in RTs or ERs across conditions because the behavioral results were similar in the two conditions.

As predicted, both conditions engaged the anterior cingulate, the insula, and prefrontal cortex (Tables 1 and 2), regions thought to be involved in interference monitoring and resolution (e.g., Carter et al., 1999; Fan, Flombaum, McCandliss, Thomas, & Posner, 2003). However, the foci of activation in the two conditions were not exactly the same. For instance, the centers of mass of the insular activation in the two conditions were 13 and 21 mm apart in the left and right hemispheres, respectively. This result suggests that there is some content (or process) domain specificity within these regions, consistent with findings such as those reported by Fan et al. (2003), in which adjacent portions of the anterior cingulate and prefrontal cortex were activated by different interference tasks (Stroop, flanker, and spatial compatibility tasks). The left hippocampus, another region that implements a rate-limiting process, showed a similar effect: Both

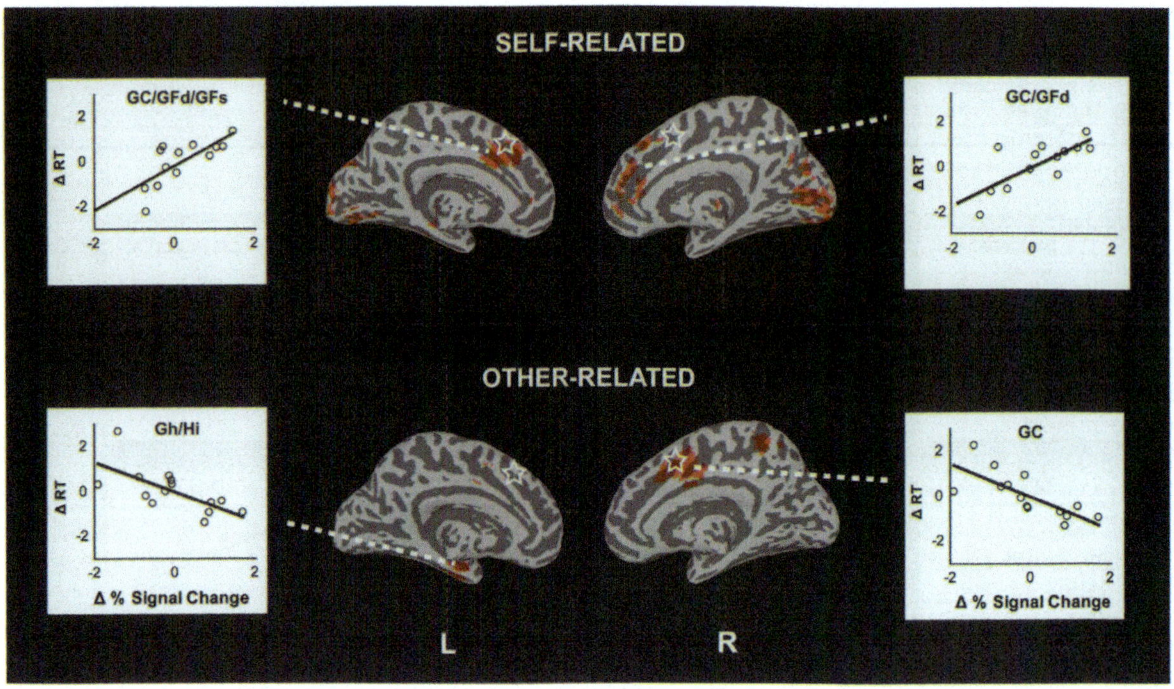

Figure 5. Medial view of the results of the main correlation analysis identifying regions that predicted the RTs across participants. Only clusters with more than 10 adjacent voxels at $p < .005$ (corresponding to $|r| > .7$) were considered significant (Tables 1 and 2). To illustrate clearly regions in which there was significant activation, the clusters shown here were drawn with a lower threshold (p = 0.01). Regions that predicted the RTs in the self-related (top) and other-related (bottom) conditions are shown in red. The stars on the cingulate and adjacent medial frontal cortex indicate the coordinates of the center of mass of the corresponding rate-limiting regions found by Wager et al. (2005) during three interference tasks. The scatterplots show RTs' z-scores as a function of brain activation z-scores for representative regions.

conditions activated the left hippocampus, but the center of mass of the focus of activation in the self-related condition was 14 mm more posterior than that in the other-related condition. The hippocampus was not a rate-limiting region in the interference tasks used by Wager et al. (2005), possibly because their tasks did not involve encoding information into, or retrieving it from, long-term memory (Schon, Hasselmo, Lopresti, Tricarico, & Stern, 2004; Svoboda et al., 2006). In contrast, such operations are necessary in both deception conditions: In our task, deceptive responses require the incidental retrieval and encoding of truthful information (as do honest responses).

A number of additional rate-limiting regions were found only in the self-related condition, including occipito-parietal and temporal regions (GOi, GOm, PCu, Cun, GL, and GF), the right thalamus, and the left cerebellum. Performance-related activation in these occipito-parietal and temporal regions may be related to visual mental imagery processes associated with retrieving autobiographical information (Ganis et al., 2003). One possibility is that individuals who produced more vivid mental images in response to the statements experienced more interference during deception. In addition, because we used written sentences as stimuli, it is likely that some of these effects were caused in part by the visual stimuli themselves, or by differential patterns of eye movements, or both. However, the fact that these regions were not found in the other-related condition suggests that an explanation based entirely on a visual stimulation artifact or on differential patterns of eye movements (e.g., more eye movements during the deceptive than honest conditions) cannot explain the pattern of results.

We also found that activation in the right thalamus predicted RTs in the Go/No-Go task used in Wager et al. (2005), which suggests that this activation may be related to motor inhibition, possibly via thalamocortical loops (Guillery & Sherman, 2002a, 2002b). Consistent with this explanation, there is single-case evidence that right thalamic dysfunction is associated with pathological lying (Modell, Mountz, & Ford, 1992). The left cerebellum activation is consistent with that found in Ganis et al. (2003), and may be reflect differences in the amount or type of memory retrieval in the two conditions (cf. Andreasen et al., 1999). Finally, a small cluster in the right inferior parietal lobule (LPi) predicted the RTs only in the other-related condition. This region may be part of the "scaffolding" network required to generate other-related lies, as discussed below.

The second result of this study is that the same cingulate–insular–prefrontal network that Wager et al. (2005) found to predict performance during three interference tasks (all of which require interference monitoring/resolution) predicted performance in the deception conditions (relative to the honest one): The centers of mass of the rate-limiting brain regions found in Wager et al. (2005) matched exactly those found in this study for the self-related condition and were very close to those found for the other-related condition. This finding provides strong evidence that these brain regions implement interference-monitoring and resolution processes that are rate-limiting not only for the interference tasks used in Wager et al. (2005), but also for the deception tasks used here. The very high correlation among these regions across individuals suggests that they may act together as a network. As in Wager et al. (2005), for self-related lies, the correlation between brain activation and the RTs was positive: Larger brain activation z-scores were associated with larger behavioral z-scores; that is, the worst liars (those who took the longest to tell a lie compared to the truth) also showed the highest brain activity during deception (compared to the truth).

Another piece of evidence that the two sorts of lies have different neural underpinnings is the fact that the correlation between brain activation and RTs had different signs in the two conditions: Unlike the self-related condition, in the other-related condition larger brain activation z-scores were associated with smaller behavioral z-scores (the worst liars showed the lowest brain activation). We can only speculate about how to interpret this unexpected result. As noted earlier, in their daily lives, people overwhelmingly lie about themselves (compared to others), as shown by studies of college students (e.g., DePaulo & Kashy, 1998). Thus, over the course of their lives most people have become practiced at lying about themselves, and practice often increases the efficiency of the neural resources required to carry out a task at the same level of performance. For instance, practice on interference tasks reduces activation in the anterior cingulate and prefrontal regions so that, after practice, it takes less neural work to achieve the same performance (e.g., Chein & Schneider, 2005; Milham, Banich, Claus, & Cohen, 2003).

Thus, the rate-limiting regions engaged during self-related lies may reflect the variability in performance among experts in this domain: The best experts have the lowest activation scores whereas the worse experts have the highest activation scores. In contrast, lying about others is a much more infrequent activity for which people may not develop expertise. Lying about others may require the effortful task of configuring a "scaffolding" brain network on the fly rather than activating an expert network already in place (Petersen, van Mier, Fiez, & Raichle, 1998). If so, then the rate-limiting regions found for other-related lies may reflect the amount of effort participants put into setting up such a network for the purpose of lying: The negative sign of the correlation for other-related lies shows that the participants with the highest brain activation scores (i.e., those who tried harder) were also the best liars.

It is important to point out some limitations of this study. First, although we used a self- vs. other-related manipulation, there are potential variables that may be correlated with this manipulation and that may be causing some of the differences. For example, there could be some differences in the emotional content of the statements used in the two conditions. Since we did not obtain ratings of the statements on emotional scales, future work will be required to address this issue. Second, the sample size ($N=14$) is rather small for an individual differences analysis. Thus, the study had limited statistical power and we may have missed important regions with relatively low correlations between brain activation and RTs.

SUMMARY

In summary, this study is a first step toward using an individual differences approach in neuroimaging research on deception. This approach specifically identifies brain regions that support rate-limiting processes. The results show that self-related and other-related lies are supported by partially nonoverlapping networks of rate-limiting regions. Of particular note, we documented that a cingulate–insular–prefrontal network, shown in other studies to be involved in interference monitoring and resolution, implements a rate-limiting process for self-related lies. A similar network was also engaged by the other-related lies; however, the activation foci were close but not overlapping with those found for the self-related lies, which suggests some regional segregation. Finally, we found that increased activation was associated with slower responses for self-related lies (relative to honest responses), but with slower responses for other-related lies (relative to honest responses). Overall, this study provides further evidence that deception is a multifaceted phenomenon, and the findings suggest that neuroimaging methods to detect deception may need to be fine-tuned for the specific types of lie to be detected. Furthermore, the findings indicate that neuroimaging studies of deception may need to take account of individual differences.

REFERENCES

Abe, N., Suzuki, M., Mori, E., Itoh, M., & Fujii, T. (2007). Deceiving others: Distinct neural responses of the prefrontal cortex and amygdala in simple fabrication and deception with social interactions. *Journal of Cognitive Neuroscience, 19*(2), 287–295.

Abe, N., Suzuki, M., Tsukiura, T., Mori, E., Yamaguchi, K., Itoh, M., et al. (2006). Dissociable roles of prefrontal and anterior cingulate cortices in deception. *Cerebral Cortex, 16*(2), 192–199.

Allen, J. J., & Iacono, W. G. (1997). A comparison of methods for the analysis of event-related potentials in deception detection. *Psychophysiology, 34*(2), 234–240.

Andreasen, N. C., O'Leary, D. S., Paradiso, S., Cizadlo, T., Arndt, S., Watkins, G. L., et al. (1999). The cerebellum plays a role in conscious episodic memory retrieval. *Human Brain Mapping, 8*(4), 226–234.

Buller, D. B., & Burgoon, J. K. (1996). Interpersonal deception theory. *Communication Theory, 3*, 203–242.

Carter, C. S., Botvinick, M. M., & Cohen, J. D. (1999). The contribution of the anterior cingulate cortex to executive processes in cognition. *Reviews in the Neurosciences, 10*(1), 49–57.

Chein, J. M., & Schneider, W. (2005). Neuroimaging studies of practice-related change: fMRI and meta-analytic evidence of a domain-general control network for learning. *Cognitive Brain Research, 25*(3), 607–623.

Collins, D., Neelin, P., Peters, T., & Evans, A. (1994). Automatic 3D intersubject registration of MR volumetric data in standardized Talairach space. *Journal of Computer Assisted Tomography, 18*(2), 192–205.

Cox, R. W. (1996). AFNI: Software for analysis and visualization of functional magnetic resonance neuroimages. *Computers and Biomedical Research, 29*(3), 162–173.

Davatzikos, C., Ruparel, K., Fan, Y., Shen, D. G., Acharyya, M., Loughead, J. W., et al. (2005). Classifying spatial patterns of brain activity with machine learning methods: Application to lie detection. *NeuroImage, 28*(3), 663–668.

DePaulo, B. M., & Kashy, D. A. (1998). Everyday lies in close and casual relationships. *Journal of Personality and Social Psychology, 74*(1), 63–79.

DePaulo, B. M., Lindsay, J. J., Malone, B. E., Muhlenbruck, L., Charlton, K., & Cooper, H. (2003). Cues to deception. *Psychological Bulletin, 129*(1), 74–118.

Fan, J., Flombaum, J. I., McCandliss, B. D., Thomas, K. M., & Posner, M. I. (2003). Cognitive and brain consequences of conflict. *NeuroImage, 18*(1), 42–57.

Gamer, M., Bauermann, T., Stoeter, P., & Vossel, G. (2007). Covariations among fMRI, skin conductance, and behavioral data during processing of concealed information. *Human Brain Mapping, 28*(12), 1287–1301.

Ganis, G., Kosslyn, S. M., Stose, S., Thompson, W. L., & Yurgelun-Todd, D. A. (2003). Neural correlates of different types of deception: An fMRI investigation. *Cerebral Cortex, 13*(8), 830–836.

Ganis, G., Thompson, W. L., & Kosslyn, S. M. (2005). Understanding the effects of task-specific practice in the brain: Insights from individual-differences analyses. *Cognitive, Affective and Behavioral Neuroscience, 5*(2), 235–245.

Guillery, R. W., & Sherman, S. M. (2002a). Thalamic relay functions and their role in corticocortical communication: Generalizations from the visual system. *Neuron, 33*(2), 163–175.

Guillery, R. W., & Sherman, S. M. (2002b). The thalamus as a monitor of motor outputs. *Philosophical Transactions of the Royal Society of London B: Biological Sciences, 357*(1428), 1809–1821.

Johnson, R., Jr., Barnhardt, J., & Zhu, J. (2004). The contribution of executive processes to deceptive responding. *Neuropsychologia, 42*(7), 878–901.

Kosslyn, S. M., Thompson, W. L., Kim, I. J., Rauch, S. L., & Alpert, N. M. (1996). Individual differences in cerebral blood flow in area 17 predict the time to evaluate visualized letters. *Journal of Cognitive Neuroscience, 8*(1), 78–82.

Kosslyn, S. M., Thompson, W. L., Shephard, J. M., Ganis, G., Bell, D., Danovitch, J., et al. (2004). Brain rCBF and performance in visual imagery tasks: Common and distinct processes. *European Journal of Cognitive Psychology, 16*, 696–716.

Kozel, F. A., Johnson, K. A., Mu, Q., Grenesko, E. L., Laken, S. J., & George, M. S. (2005). Detecting deception using functional magnetic resonance imaging. *Biological Psychiatry, 58*(8), 605–613.

Kozel, F. A., Padgett, T. M., & George, M. S. (2004). A replication study of the neural correlates of deception. *Behavioral Neuroscience, 118*(4), 852–856.

Kruggel, F., & von Cramon, D. Y. (1999). Temporal properties of the hemodynamic response in functional MRI. *Human Brain Mapping, 8*(4), 259–271.

Langleben, D. D., Loughead, J. W., Bilker, W. B., Ruparel, K., Childress, A. R., Busch, S. I., et al. (2005). Telling truth from lie in individual subjects with fast event-related fMRI. *Human Brain Mapping, 26*(4), 262–272.

Langleben, D. D., Schroeder, L., Maldjian, J. A., Gur, R. C., McDonald, S., Ragland, J. D., et al. (2002). Brain activity during simulated deception: An event-related functional magnetic resonance study. *NeuroImage, 15*(3), 727–732.

Lee, T. M. C., Liu, H. L., Chan, C. C., Ng, Y. B., Fox, P. T., & Gao, J. H. (2005). Neural correlates of feigned memory impairment. *NeuroImage, 28*(2), 305–313.

Lee, T. M. C., Liu, H.-L., Tan, L.-H., Chan, C. C. H., Mahankali, S., Feng, C.-M., et al. (2002). Lie detection by functional magnetic resonance imaging. *Human Brain Mapping, 15*, 157–164.

Macwhinney, B., Cohen, J., & Provost, J. (1997). The PsyScope experiment-building system. *Spatial Vision, 11*(1), 99–101.

Maguire, E. A. (2001). Neuroimaging studies of autobiographical event memory. *Philosophical Transactions of the Royal Society of London B: Biological Sciences, 356*(1413), 1441–1451.

McCornack, S. A. (1992). Information manipulation theory. *Communication Monographs, 59*, 1–16.

McCornack, S. A. (1997). The generation of deceptive messages: Laying the groundwork for a viable theory of interpersonal deception. In J. O. Greene (Ed.), *Advances in Communication Theory* (pp. 91–126). Mahwah, NJ: Lawrence Erlbaum.

Milham, M. P., Banich, M. T., Claus, E. D., & Cohen, N. J. (2003). Practice-related effects demonstrate complementary roles of anterior cingulate and prefrontal cortices in attentional control. *NeuroImage, 18*(2), 483–493.

Miller, M. B., Van Horn, J. D., Wolford, G. L., Handy, T. C., Valsangkar-Smyth, M., Inati, S., et al. (2002). Extensive individual differences in brain activations associated with episodic retrieval are reliable over time. *Journal of Cognitive Neuroscience, 14*(8), 1200–1214.

Modell, J. G., Mountz, J. M., & Ford, C. V. (1992). Pathological lying associated with thalamic dysfunction demonstrated by [99mTc]HMPAO SPECT. *Journal of Neuropsychiatry and Clinical Neuroscience, 4*(4), 442–446.

Mohamed, F. B., Faro, S. H., Gordon, N. J., Platek, S. M., Ahmad, H., & Williams, J. M. (2006). Brain mapping of deception and truth telling about an ecologically valid situation: Functional MR imaging and polygraph investigation—Initial experience. *Radiology, 238*(2), 679–688.

Morgan, C. J., Tolley, J. B., & Kosslyn, S. M. (in press). Types of deception revealed by individual differences in cognitive abilities. *Social Neuroscience.*

National Research Council. (2003). *The polygraph and lie detection*. Washington, DC: National Academies Press.

Ng, V. W., Bullmore, E. T., de Zubicaray, G. I., Cooper, A., Suckling, J., & Williams, S. C. (2001). Identifying rate-limiting nodes in large-scale cortical networks for visuospatial processing: An illustration using fMRI. *Journal of Cognitive Neuroscience, 13*(4), 537–545.

Nunez, J. M., Casey, B. J., Egner, T., Hare, T., & Hirsch, J. (2005). Intentional false responding shares neural substrates with response conflict and cognitive control. *NeuroImage, 25*(1), 267–277.

Oldfield, R. C. (1971). The assessment and analysis of handedness: The Edinburgh inventory. *Neuropsychologia, 9*(1), 97–113.

Petersen, S. E., van Mier, H., Fiez, J. A., & Raichle, M. E. (1998). The effects of practice on the functional anatomy of task performance. *Proceedings of the National Academy of Sciences of the United States of America, 95*(3), 853–860.

Phan, K. L., Magalhaes, A., Ziemlewicz, T. J., Fitzgerald, D. A., Green, C., & Smith, W. (2005). Neural correlates of telling lies: A functional magnetic resonance imaging study at 4 Tesla. *Academic Radiology, 12*(2), 164–172.

Plomin, R., & Kosslyn, S. M. (2001). Genes, brain and cognition. *Nature Neuroscience, 4*(12), 1153–1154.

Roediger, H. L., & Marsh, E. J. (2003). Episodic and autobiographical memory. In A. F. Healy & R. W. Proctor (Eds.), *Experimental psychology* (Vol. 4 pp. 475–497). New York: Wiley.

Schaefer, A., Braver, T. S., Reynolds, J. R., Burgess, G. C., Yarkoni, T., & Gray, J. R. (2006). Individual differences in amygdala activity predict response speed during working memory. *Journal of Neuroscience, 26*(40), 10120–10128.

Schon, K., Hasselmo, M. E., Lopresti, M. L., Tricarico, M. D., & Stern, C. E. (2004). Persistence of parahippocampal representation in the absence of stimulus input enhances long-term encoding: A functional magnetic resonance imaging study of subsequent memory after a delayed match-to-sample task. *Journal of Neuroscience, 24*(49), 11088–11097.

Sidtis, J. J. (2007). Some problems for representations of brain organization based on activation in functional imaging. *Brain and Language, 102*(2), 130–140.

Spence, S. A., Farrow, T. F., Herford, A. E., Wilkinson, I. D., Zheng, Y., & Woodruff, P. W. (2001). Behavioural and functional anatomical correlates of deception in humans. *NeuroReport, 12*(13), 2849–2853.

Spence, S. A., Hunter, M. D., Farrow, T. F., Green, R. D., Leung, D. H., Hughes, C. J., et al. (2004). A cognitive neurobiological account of deception: Evidence from functional neuroimaging. *Philosophical Transactions of the Royal Society of London B: Biological Sciences, 359*(1451), 1755–1762.

StataCorp (2001). *Statistical Software: Release 7.0.* College Station, TX: Stata Corporation.

Svoboda, E., McKinnon, M. C., & Levine, B. (2006). The functional neuroanatomy of autobiographical memory: A meta-analysis. *Neuropsychologia, 44*(12), 2189–2208.

Tulving, E. (2002). Episodic memory: From mind to brain. *Annual Review of Psychology, 53*, 1–25.

Vrij, A. (2000). *Detecting lies and deceit.* Chichester, UK: Wiley.

Wager, T. D., Sylvester, C. Y., Lacey, S. C., Nee, D. E., Franklin, M., & Jonides, J. (2005). Common and unique components of response inhibition revealed by fMRI. *NeuroImage*, *27*(2), 323–340.

Xiong, J., Gao, J.-H., Lancaster, J. L., & Fox, P. T. (1995). Clustered analysis for functional MRI activation studies of the human brain. *Human Brain Mapping*, *3*, 287–301.

Zuckerman, M., DePaulo, B. M., & Rosenthal, R. (1981). Verbal and nonverbal communication of deception. In L. Berkowitz (Ed.), *Advances in experimental social psychology* (Vol. 14 pp. 1–59). New York: Academic Press.

SOCIAL NEUROSCIENCE, 2009, 4 (6), 554–569

Psychology Press
Taylor & Francis Group

Types of deception revealed by individual differences in cognitive abilities

Charity J. Morgan, Julia B. LeSage, and Stephen M. Kosslyn
Harvard University, Cambridge, Massachusetts, USA

The two studies reported in this article are an extension of the neuroimaging study by Ganis et al. (2003), which provided evidence that different types of lies arise from different cognitive processes. We examined the initial response times (IRTs) to questions answered both deceptively and truthfully. We considered four types of deceptive responses: a coherent set of rehearsed, memorized lies about a life experience; a coherent set of lies spontaneously created about a life experience; a set of isolated lies involving self-knowledge; and a set of isolated lies involving knowledge of another person. We assessed the difference between truthful and deceptive IRTs. Scores from cognitive tasks included in the MiniCog Rapid Assessment Battery (MRAB) were significant predictors of IRT differences. Each type of lie was predicted by a distinct set of MRAB scores. These results provide further evidence that deception is a multifaceted process and that different kinds of lies arise from the operation of different cognitive processes.

INTRODUCTION

Deception involves intentionally misleading another person, either by omission or commission. For both theoretical and practical reasons, many researchers have studied how to detect deception, from numerous vantage points, over the course of many years (Burgoon & Buller, 1994; Ekman, 1992, 2001; Horvath, Jayne, & Buckley, 1994; Mehrabian, 1971; Vrij, 1994; Zuckerman, DeFrank, Hall, Larrance, & Rosenthal, 1979). Nevertheless, to date no method of detecting deception has proven satisfactory (e.g., National Research Council, 2003). In particular, the standard approach, based on measuring signals at the periphery (such as skin conductance, respiration rate, blood pressure, and heart rate), has been highly criticized. Polygraph tests rely on the assumption that deception creates arousing emotions such as guilt or fear, and such tests are designed to detect the physiological changes that occur during deception. Such changes are measured at the periphery, and typically rely on assessing heart rate, skin conductance, and blood pressure. However, not only has no unique physiological pattern been found to be associated with deception (Iacono, 2000), but also many researchers even question the validity of the assumptions on which the polygraph is based (Iacono & Lykken, 1997). Furthermore, the results of laboratory studies on the validity of the polygraph may not be generalizable to the real world—and the accuracy of polygraph results in the field is difficult to determine because

Correspondence should be addressed to: Charity Morgan, Department of Statistics, One Oxford Street, Cambridge, MA 02138, USA. Email: cmorgan@stat.harvard.edu

Julia B. LeSage is now in Macon, Georgia. This material is based partially on work supported by the National Science Foundation under Grant BCS-0322611. Any opinions, findings, and conclusions or recommendations expressed in this material are those of the authors and do not necessarily reflect the views of the National Science Foundation. The authors wish to thank Jennifer M. Shepard, William L. Thompson, and Giorgio Ganis for their helpful advice and constructive criticism.

www.psypress.com/socialneuroscience DOI:10.1080/17470910802299987

ground truth cannot always be established (Vrij, 2008).

In response to the drawbacks of detecting deception by registering signals at the periphery, researchers have turned to studying the brain events that underlie deception. Two broad approaches have emerged. On one hand, some researchers have sought to discover a "neural signature" of deception. For example, studies have shown that the anterior cingulate—a brain area thought to be involved in monitoring conflict between observed and expected events—is active when one engages in deception (Kozel et al., 2004; Langleben et al., 2002). On the other hand, Ganis, Kosslyn, Stose, Thompson, and Yurgelun-Todd (2003) suggested that there is no single neural signature of deception. Instead, these researchers argue that there are numerous different ways to lie, and different neural networks are evoked depending on the type of lie. In particular, Ganis et al. (2003) showed that different brain areas are activated when a person lies spontaneously than when a person lies on the basis of a previously memorized scenario.

The present research follows up the Ganis et al. (2003) functional magnetic resonance imaging (fMRI) study, using an entirely different logic to document that different processes can be used to tell lies. In these studies we exploit the fact that people differ in how effectively they can perform specific information processing, such as monitoring conflict or retrieving stored information. If different processes are evoked to produce different sorts of lies, then individual differences in the efficacy of different processes should predict performance when participants produce different types of lies. We use this logic to extend the inferences drawn by Ganis et al., and document an additional dimension that underlies the production of deception.

The present studies relied on initial response times (IRTs), which were defined as the length of time between the end of a question and the beginning of a response. Greene, O'Hair, Cody, and Yen (1985) characterized an increased latency between question and response as "indicative of heavy demands upon the central processing capacity" and reasoned that longer latencies when providing a deceptive response would be consistent with the theory that "lying generally requires more cognitive work than telling the truth." Cody, Marston, and Foster (1984) also consider the IRT to be a reliable indicator of "the cognitive effort required for successful deception [that] cause[s] communicators to exhibit longer response latencies when responding deceptively ... than when responding truthfully" (as cited in Stiff, Corman, Krizek, & Snider, 1994). Seymour, Seifert, Shafto, and Mosmann (2000) found that IRT could reliably distinguish which participants in their study were in possession of "guilty knowledge." Similarly, Farrow et al. (2003) asked participants questions about themselves and found that participants required longer IRTs when lying than when telling the truth. Sporer and Schwandt's (2006) meta-analytic study concluded that people tend to take slightly longer to respond to questions when lying than when telling the truth.

However, telling a lie may not always be signaled by increased IRT. McCornack (1997) notes that the cognitive effort required to produce a deceptive message can vary. We propose that such variations could reflect differences in the type of lie being produced as well as individual differences in the deceiver's cognitive abilities. It comes as no surprise, then, that a number of studies have found that participants do not always have longer IRTs during deception. For example, Greene et al. (1985) found that people actually responded more quickly with a prepared lie than they did with the truth. Greene et al. did not find any group differences when the lies told were spontaneous. Vrij (2008) summarizes several studies that consider IRT and finds an inconsistent relationship between the length of IRT and deception.

One limitation of previous research on deception is that investigators often ignore or underestimate the importance of the ways in which lies can differ. For example, deTurck and Miller (1985) designed an experiment in which participants cheated on a task performed with a confederate. Participants who subsequently tried to conceal their cheating therefore had to lie not only about their own behavior but also about the behavior of the confederate. The fact that some of the participants' deceptive statements concerned their own behavior and others concerned the behavior of the confederate is a potentially confounding factor that was not examined nor acknowledged. In contrast, Greene and colleagues (1985) examined prepared lies and spontaneous lies separately, and hypothesized that creating a lie spontaneously would require more cognitive effort than reciting a lie that had been prepared beforehand. However, they did not investigate possible differences in cognitive demands that might underlie different sorts of lies.

Another limitation of previous deception research is the lack of focus on individual differences, particularly cognitive differences. For example, Ganis et al. (2003) and Greene et al. (1985) report group differences in response time and do not compare IRT within individuals. Although Farrow et al. (2003) do consider individual differences in IRT, they cast these differences solely in the context of sex and personality traits. In the studies reported here, we compared IRT for telling the truth versus lying for each individual, essentially using each participant as his or her own control. Furthermore, we examine the relationship between an individual's IRT differences and his or her cognitive abilities.

IRT differences can also be useful in judging whether an individual is skilled at deception. An individual's attempt to deceive another can only truly be judged successful if the recipient of the deceptive message is indeed fooled by it. In the studies reported here, we focus only on the deceivers and therefore cannot judge the effectiveness of the participants' attempts at deception in this way. Instead we reason that in order for attempts at deception to be successful, a lie must be indistinguishable from the truth. One of several possible ways that a deceptive response can differ from a truthful one is in the amount of time required to produce it. We expect that some successful deceivers will require approximately the same amount of time to tell a lie as to tell the truth; that is, effective deception should result in IRT difference scores close to zero. However, the ability to deceive effectively should depend, at least in part, on how well a person can manifest relevant cognitive abilities. Thus, we expected that the difference in IRTs between truthful and deceptive responses should vary depending on the type of lie and on the relevant cognitive strengths and weaknesses of the participant.

To assess individual differences in cognitive abilities, we used the MiniCog Rapid Assessment Battery (MRAB, Shephard & Kosslyn, 2005). This battery consists of nine tasks, which rely upon response time and error rate measurements. These tasks assess three sorts of attention (selective, divided, and vigilance), two kinds of working memory (verbal and spatial), three kinds of reasoning (three-term series verbal deduction, spatial mental rotation, and set-switching), and simple perceptual-motor reaction time. A summary of these tasks is provided in the appendix. If different sets of individual differences predict IRTs for different types of lies, that is evidence that different sorts of processing underlie the production of different types of lies.

EXPERIMENT 1

This experiment essentially replicated the behavioral methods of the Ganis et al. (2003) fMRI study. We compared lies when participants lied spontaneously versus when they lied based on a previously memorized story. However, instead of assessing patterns of neural activity, we measured individual differences on the MRAB in advance, and then measured IRT and error rates (ERs) during the production of deception and truthful responses.

Based on the earlier fMRI results, we had one strong prediction: Individual differences on the MRAB variant of the classic Stroop task—a "filtering" task that required indicating how many repetitions of a digit were present, independent of what the digit actually was—should predict the results when participants lied spontaneously but not when they lied on the basis of a memorized story. Ganis et al. (2003) found consistent activation of the anterior cingulate only during the spontaneous condition, and this structure is strongly activated during variants of the Stroop task (Bush et al., 1998).

Method

Participants

Thirty-nine participants (20 females and 19 males) volunteered to take part in the study for pay. Five participants (four females and one male) were removed from the sample because they had unusually high error rates (13.64% or greater) on the computer-administered deception task. The final sample consisted of 16 females (mean age 23.5 years; range 19–31 years) and 18 males (mean age 23.44 years; range 18–35 years). Seven females and 11 males were Harvard students, and the remaining participants were from the community or other local colleges. The mean for the Bors and Stokes (1998) short form (12-item) of Raven's Advanced Progressive Matrices (APM) for the females was 8.06 (range 2–12) and for the males was 7.78 (range 3–12). The mean number of years of education for the females was 16.25 (range 12–22) and for the males was 15.44 (range 12–20); 14 females and 14 males were right-handed. All participants were US citizens or permanent residents, had lived in the USA for at least the past 10 years, spoke fluent English, did not stutter, and were physically and psychologically healthy. The fliers and studypool information used to recruit participants did not mention that the study was designed to investigate the nature of deception, but instead described it only as a "cognitive" study. The study was conducted with the informed consent of each participant and the approval of the university's Institutional Review Board.

Materials

A critical part of this study, and the most time-consuming, was the interaction between the investigator and the participant *before* the participant arrived in the lab. The investigator sent participants an email message, asking them to review their most memorable work experience and their most memorable vacation experience. The participant was asked to write short summaries of these events, and to send the descriptions to the investigator, along with a phone number, and a time the participant could come into the lab. Each of the two experiences was to be summarized in one page, which was to include a detailed description (who, what, when, where, and why) of the event. If the descriptions did not include enough detail, the investigator asked for more detail, but this had to be done only twice (both were cases in which the participants sent only half a page instead of a full page for each event).

The investigator generated questions from the descriptions, and called the participant the day before the scheduled session (which was an average of one week after the descriptions were received). During this phone call, the participant was instructed to invent an alternate reality, a vacation or work experience (the selection of work or vacation alternated with each participant) that never happened but could have happened. The investigator wrote down everything the participant said, while at the same time, coaching the participant to include details that the investigator needed for the prepared questions. We used the same procedure as Ganis et al. (2003, p. 832):

> For example, if they actually took their vacation in Florida, we instructed them to pretend that the vacation took place in another location in the United States (e.g. California); if they traveled there by car, we instructed them to pretend they used another means of transportation (e.g. by plane), and so on. We helped the participants in this process by ensuring that the scenarios they generated were coherent and internally consistent.

However, in this study, the investigator explicitly asked the participant to think of a *plausible* alternative, such as another vacation that he/she was actually considering. Most of the participants could easily think of the place where they could have gone on vacation or another plausible event at work. The investigator probed for details about that alternative, but never actually posed the questions that were used in the study. We created a transcript of this "alternative reality." Approximately 30 min after this conversation, the investigator emailed the participant the transcript of the alternative reality to memorize *before* they went to sleep that night. We asked them to visualize themselves going on this vacation or working at this place, which would help them to memorize the details of the scenario. We also told the participants that they had to memorize the scenario in detail because they would be asked questions about it when they came into the lab.

We created two sets of 11 questions for each participant, one set for the work scenario and one set for the vacation scenario. Each of these sets was presented to the participant twice, with one presentation requiring true responses and one requiring deceptive responses. The "alternative reality" scenario was used to produce the deceptive responses for the relevant event for a given participant. All questions were recorded on the computer prior to the participant's arriving at the lab, using SoundEdit 16 software. Unlike Ganis et al. (2003), we also avoided asking questions about the time period of the event, because we wanted the alternative reality to be plausible even if it was not what actually occurred at that point in time.

Procedure

The participant came to the lab and completed the informed consent form. After this informed consent had been completed, we administered the nine MRAB tasks, presented on a Palm Tungsten C personal digital assistant (PDA). The investigator began by reading a prepared script, which explained to each participant how to use the PDA, how to position the body in the chair, how to position the thumbs over the PDA buttons, and how to proceed through the nine tasks. The script ended by instructing the participants to complete each of the nine tasks as quickly and as accurately as possible. The investigator remained in the room, behind and out of sight of the participant, for the entire study and answered any questions the participants may have had.

Once the participant completed the MRAB, the investigator sat across from the participant and asked whether he or she had memorized the "alternative reality" scenario before going to sleep the previous night. Each participant said that he or she had memorized the alternate scenario. Next, the investigator asked the participant to recite this scenario. If the participant missed a detail, the investigator helped the participant to think about that detail, without asking the participant a question. It was rare that the participant did not mention all the details; when that did occur, the participant quickly remembered upon being told that he or she needed to talk more about a certain aspect of the scenario. Finally, the investigator asked the participant to review and then discuss the written transcripts of both of the true scenarios and the single false alternate scenario.

The investigator then instructed the participant to move close to the microphone (which was placed on a stand on the desk). A video camera was attached to the top of the computer and the keyboard was off to the side of the computer. The participant read the instructions on the computer screen, pressing the space bar to progress to the next page on the screen. The instructions indicated that a second investigator would be watching the participant's face from another room, and would be trying to determine on which of the four blocks the participant had lied. Participants were told that every time the observer detected lies, the participant would lose $4 from the hourly wage; every time the participant succeeded in deceiving the observer, he or she would gain $4. When the participant came to the page that described the video camera and the fact that he/she was being watched, the investigator turned the camera on, which produced a clearly audible "beep" so that the participant knew that it was operating. The investigator would say, "You are now being watched." The investigator would then leave the room briefly and go across the hall, pretending that she was checking with the "second lab member" who was viewing the participant. She would come back in the room and say, "You are positioned correctly. He can see your face," so that the participants would really believe that they were being observed. Although participants were told that their earnings depended in part on successfully deceiving the observer, in fact there was no lab member watching the experiment and all participants were paid the full amount as if they had successfully deceived the observer.

The investigator next described the individual trials. The cue at the beginning of each block indicated which event would be probed, and how the participant should respond. For example, one cue read "Please answer the following questions about your VACATION experience with FALSE answers." Participants were asked to respond to each question with a one word response. They were asked not to say "um," and to remain silent between questions.

The last screen of the instructions asked the participant to paraphrase what he/she would be doing during the study. The investigator would not proceed to the practice trials until this paraphrase was completely correct. For example, the instructions explained that there would be four blocks randomly presented. One would be a block of questions for which the participants had to give true answers about their vacation, one would require false answers about their vacation, one would require true answers about their work, and one would require false answers about their work. If the participants had a memorized, alternate scenario for their work memory, the investigator would again explain that the false answers for the work memory had to come from the alternative reality, and nothing else. For the other event, which did not have a previously formulated alternative reality, the investigator went on to explain that the spontaneous false answers had to make sense. For example, if the participant said he went to Hawaii, he could not say that he saw a polar bear there—and, the participant was reminded, such a response included two words, but an appropriate response should include only a single word.

Once the investigator was satisfied that the participants understood the procedure, participants were allowed to proceed to the practice trials. The final instruction on the screen urged the participants to "Please respond as quickly and as accurately as possible."

Participants were presented with four practice questions, one for each block of instructions.

Before each practice question, participants were given a cue that indicated how they were to answer the following question, and were asked to push the space bar to continue. Following the presentation of the cue, a question was presented auditorially. Once the participant had responded, a fixation point appeared on the screen to signal that the next cue would appear shortly. The fixation point remained on the screen for 1 s. After the practice trials the investigator talked with the participant about any errors in the responses. For example, if the participant said "um," it was caught during the practice trials and the instructions were reiterated by the investigator.

The participants were then presented with four blocks of questions, with each block containing 11 questions, for a total of 44 questions. To stay as close to the Ganis et al. (2003) procedure as possible, the truth blocks came after the lie blocks, but the order of the vacation or work blocks was counterbalanced. All the questions in the blocks were presented in a random order. We presented both lie conditions first in order to avoid potential short-term interference from the truth conditions (such as actively having to inhibit primed responses; see Ganis et al., 2003).

All questions were presented auditorially by the PsyScope program (Cohen, MacWhinney, Flatt, & Provost, 1993), on a Macintosh computer. Upon presentation of the question, the IRT of the participant's spoken reply was registered by the PsyScope program, via a microphone attached to the computer.

Once the actual experimental trials began, the investigator sat behind the participant so that she could see the computer screen and the dot that appeared instantly when the participant made a response. She recorded the verbal responses at the same time they were made and recorded any responses that were invalid (two words instead of one, "um," or deviation from the alternative reality scenario). Error rates later were calculated from this information.

After the four blocks of test trials, the investigator shut down the camera. If the investigator had any questions about the responses, she asked the participant at this time. Finally, we administered the Bors and Stokes (1998) short form of the Raven's APM, along with a debriefing form that asked not only whether participants had any problems during the study, but also questions pertaining to MRAB and biorhythms, such as "What time did you go to sleep last night?" and "Did you have breakfast this morning?". These questions were asked in order to analyze the MRAB data in other, larger databases.

Results

For each of the four conditions (Work/Truth; Work/Lie; Vacation/Truth; Vacation/Lie, where one of the lie conditions was spontaneous and one memorized), we calculated difference scores for each participant by subtracting the mean IRT for the valid truthful responses from the mean IRT for the valid deceptive responses (we did not include IRTs for trials on which participants made errors). We grouped the deceptive responses according to whether the lies were spontaneous or based on the memorized alternative reality.

We began with a two-way analysis of variance, comparing the difference scores for Spontaneous versus Memorized lies for men versus women. There was no effect of gender, $F<1$, nor an interaction between gender and condition (Wilks' $\lambda=.96$, $F=1.39$, $p=.24$). Thus, in all subsequent analyses we pooled over gender.

In the Memorized condition, the participants tended to take less time to lie than to tell the truth (with means of 1345 and 1523 ms for the lie and truth conditions, respectively), $t(33)=-1.78$, $p=.08$ (two-tailed). This result is consistent with Greene and colleagues' (1985) finding that participants responded faster when they provided a memorized lie than when they provided the truth. We speculate that participants may have had easier access to the lie than to the truth (which was not as well rehearsed) following the recent rehearsal and intense emphasis on memorization of the alternate scenario; that is, reciting the well-memorized lie may have required less cognitive effort than telling the truth, resulting in lower IRTs.

In contrast, in the Spontaneous condition we did not find even a tendency toward a difference between the two conditions (with means of 1667 and 1599 ms for the lie and truth conditions, respectively), $t<1$, $p>.1$ (two-tailed). This result also agrees with the findings of Greene et al. (1985), who reported no difference between the response times for participants providing a spontaneous lie and those providing the truth. For the Spontaneous condition, the difference between the cognitive processes required to produce a spontaneous lie and those required to provide a truthful response may not be captured by IRT differences alone. In addition, despite the fact that the alternate scenario was not used in the Spontaneous condition, perhaps having a well memorized story available interfered with participants' performance in the Spontaneous condition. Alternatively, perhaps the participants were fatigued by the time the truth blocks were presented, although there was no evidence of such an effect in Ganis et al. (2003).

Next, for each of the nine MRAB tasks, we calculated each participant's average response time (RT) and error rate (ER). All RT calculations were based only on the trials that the participant answered correctly. We hypothesized not only that the MRAB cognitive measures can predict performance on deception tasks, but also that different subsets of these measures will predict performance for different types of lies. We also included each participant's score on the shortened Raven's form as a means of controlling for general intelligence.

Next, we used a forward stepwise multiple regression analysis to examine the independent contributions of the different MRAB scores, if any, for predicting the IRT difference scores.

The mean Raven's score for the participants is significantly higher ($p = .04$) than the norms reported in Bors and Stokes (1998). In order to examine the impact of the participants' relatively high Raven's scores, we analyzed the data both with and without using the Raven's score as a predictor variable. For the first two stepwise regression analyses performed on data from each condition, we forced the short-form Raven's score in as a predictor at the outset; that is, we selected the Raven's score as a predictor of the IRT difference scores, and allowed the stepwise regression procedure to determine which MRAB scores predicted IRT difference scores after controlling for the Raven's scores. In the second pair of regression analyses, we omitted the Raven's score altogether as an independent variable, and only included the MRAB scores. The stepwise regression procedure then determined which MRAB scores predicted the IRT difference scores without controlling for general intelligence. The results of the stepwise regressions are summarized in Tables 1 and 2.

First, we note that the results change depending on whether the Raven's score is forced into or excluded from the regression analysis. However, in either case, the crucial result stands: Specifically, we found that different MRAB results predict the IRT difference scores in the Spontaneous versus Memorized conditions.

In order to document further that individual differences in different processes predict specific types of lies, we asked whether either of the sets of predictors identified by the stepwise regression analysis could predict both types of lies successfully. Each IRT difference score was fit to both sets of MRAB results, with the Raven's score excluded from the analyses. To account for the fact that the two types of lies have differing numbers of MRAB predictors, for each fit we calculated the Akaike Information Criterion (AIC; Akaike, 1974). A better fit can always be obtained by adding more predictors; the AIC evaluates the fit of a linear model by using maximum likelihood, but also adds a penalty based on the number of predictors used in the model. When comparing linear models with the same response variable, the model with the lower AIC value fits the data better. The results are presented in Table 3. The set of predictors identified by the stepwise regression in the Memorized condition (AIC = 524) explains the IRT difference scores for memorized lies better

TABLE 1
Stepwise regression analysis for Experiment 1, with IRT difference scores as criteria and MRAB scores as predictors: Raven's scores forced in at the outset

Criteria	*Predictors*	*β*	*Partial R^2*	*Model R^2*	*F*
Memorized	Raven's score	0.31	.17	.17	6.24**
	Mental rotation RT	−0.51	.13	.30	5.50**
	Vigilance RT	0.36	.11	.41	5.17**
Spontaneous	Raven's score	−0.05	.01	.01	0.22
	Spatial working memory RT	−0.77	.09	.10	2.93*
	Verbal working memory RT	0.80	.12	.22	4.36**
	Vigilance ER	0.50	.10	.32	3.96*
	Filtering RT	−0.36	.11	.43	5.12**

Notes: *$p < .10$; **$p < .05$; ***$p < .01$; ****$p < .005$.

TABLE 2
Stepwise regression analysis for Experiment 1, with IRT difference scores as criteria and MRAB scores as Predictors: Raven's scores not included

Criteria	*Predictors*	*β*	*Partial R^2*	*Model R^2*	*F*
Memorized	Mental Rotation RT	−0.50	.22	.22	8.93***
	Vigilance RT	0.37	.09	.31	4.11*
	Cognitive set switching ER	−0.31	.07	.38	3.14*
	Spatial working memory RT	−0.25	.05	.43	2.48
Spontaneous	Spatial working memory RT	−0.78	.09	.09	2.97*
	Verbal working memory RT	0.80	.11	.20	3.97*
	Vigilance ER	0.52	.12	.32	4.92**
	Filtering RT	−0.36	.11	.43	5.21**

Notes: *p <.10. **p <.05. ***p <.01. ****p <.005.

than those identified in the Spontaneous condition (AIC = 538). The reverse is true for the Spontaneous IRT difference scores: The set of predictors identified in the Spontaneous condition (AIC = 526) explains the IRT difference scores for spontaneous lies better than the set of predictors identified in the Memorized condition (AIC = 539). Thus, we see that no single set of MRAB scores better predicts the IRT scores for both types of lies.

Discussion

The results revealed that different individual differences predict the IRT measure (the difference in time to initiate a lie versus to tell the truth) for Memorized versus Spontaneous deceptive responses. In particular, we see that Filtering RT (the MRAB Stroop task) was identified as a significant predictor for the Spontaneous, but not the Memorized, IRT difference scores. This finding is evident regardless of whether or not the Raven's scores are included in the equations.

The Memorized lie model estimated with the Raven's score excluded from the analyses explains a higher percentage of the variance than when the Raven's score is forced into the regression (and for the Spontaneous lie models, the Raven's simply doesn't contribute significantly to the variance), so we discuss the models that do not include the Raven's in further detail.

TABLE 3
AIC for IRT difference scores fitted against identified sets of MRAB results for Experiment 1

	IRT Difference scores	
MRAB results	*Memorized*	*Spontaneous*
Mental rotation RT, vigilance RT, cognitive set switching ER, spatial working memory RT	524	539
Spatial working memory RT, verbal working memory RT, vigilance ER, filtering RT	538	526

For the Memorized IRT difference scores, RT on the MRAB Mental Rotation task accounts for the highest portion of the variance explained by the model. Mental Rotation RT was not identified as a significant predictor of the Spontaneous IRT difference scores. However, contrary to what we might have expected, Mental Rotation RT is negatively associated with the Memorized IRT difference scores. That is, participants with worse mental imagery—as reflected in the Mental Rotation scores—had a larger negative difference between the amount of time required to produce lies as compared to the truth. Our speculation (and it is only that) is that the participants may have used mental imagery to provide truthful responses to the questions. Perhaps this mental imagery was not needed to produce the lies in the Memorized condition as those lies were well rehearsed. Along these lines, this mental imagery would have been necessary to produce both the truth and lies in the Spontaneous condition, which could explain why Mental Rotation RT did not significantly predict the difference between lies and truth in that condition. We stress that this is only a post-hoc speculation and whatever the reason for this result, the crucial finding is that a different set of MRAB tasks was identified for the Memorized IRT difference scores than for the Spontaneous IRT difference scores.

For the Spontaneous IRT difference scores, ER on the MRAB Vigilance task accounts for the highest portion of the variance explained by the model. Vigilance ER was not identified as a significant predictor of the Memorized IRT difference scores. Higher ER on this task is associated with larger differences between truth and lie IRTs, and thus Vigilance RT is related to the participants' responding faster with the truth or more slowly with lies. It is possible that vigilance reflects the participants' being poised for the cue, and ready to respond to the truth; if so, they would be faster if they were more vigilant to the cue and slower if they had to overcome such priming. But again, this is merely a post-hoc account.

We stress that the key finding here is that individual differences in different processes assessed by MRAB predicted the time to produce the different sorts of lies. This finding is *prima facie* evidence that different processes are used in the two conditions. Moreover, the fact that our version of the Stroop task ("Filtering") predicted times only in the Spontaneous condition, as expected on the basis of the fMRI results, can be treated as a form of validation of the methodology.

EXPERIMENT 2

Experiment 1 served to show that a novel behavioral method could be used to document that different sorts of lies rely on different underlying processes. In this experiment we use this method to chart new territory, to ask whether lies could differ along another dimension. There is ample evidence that the "self" is represented differently in the brain than "others." For example, words about the self are processed by different brain regions than words about others (Craik et al., 1999; Gusnard, Akbudak, Shulman, & Raichle, 2001; Kelley et al., 2002). Moreover, it is well known that the monkey brain has distinct regions for representing space around the self versus other regions of space (Berti & Frassinetti, 2000; Gross & Graziano, 1995).

Thus, in this experiment we used the same individual-differences approach to investigate whether there is a difference between lying about oneself versus lying about another person. If in fact the results document this distinction, we then can compare these findings to those from the previous experiment. It is possible that there are at least four systems that underlie deception. All of the responses in this study were spontaneous, but it is possible that spontaneous lies about facts regarding the self or others (which may be essentially semantic memories) both differ from spontaneous lies about episodic memories of events.

Method

Participants

Forty-two participants (22 females and 20 males) volunteered to participate for pay. Two participants (one female and one male) were removed from the sample for failing to meet the study prerequisites. Seven participants (five females and two males) were removed from the sample for having fewer than seven valid responses for at least one of the four conditions. A response given by a participant was considered valid if, for that question, the participant both adhered to the cue and knew the truthful response to the question. The final sample consisted of 16 females (mean age 21.7 years; range 18–33 years) and 17 males (mean age 23.6 years; range 18–35 years). Eleven females and 12 males were Harvard students, and the remaining participants were from the community or other local colleges. None of these people had participated in the previous experiment. The mean score on the Bors and Stokes (1998) short form of the Raven's APM for the females was 8.63 (range 1–12) and for the males was 8.06 (4–12). The mean number of years of education for the females was 14.25 years (range 12–21 years) and for the males was 14.06 years (range 12–18 years); 12 females and 16 males were right-handed. The inclusion criteria noted in Experiment 1 were also applied in Experiment 2.

Materials

We prepared two sets of 24 questions; one set required a single-word response about oneself (used in the "Self" condition), and the other set required a single-word response about President George W. Bush (used in the "Other" condition). President Bush was chosen to be the subject of the "Other" questions because we assumed that most participants would know a wide range of information about him. The questions asked the participants about factual conditions. For example, two questions were "What is your home state?" and "How many children does Bush have?" The questions were paired in that, for each fact, participants were asked both about themselves and about President Bush. For example, participants were asked both "What is your job title?" and "What is Bush's job title?".

However, once the study had been completed but before performing the data analysis, we decided that five of the "Self" questions did not strictly ask about the participants themselves. We eliminated these questions because each of these items asked the participants to provide information not strictly about themselves but about a person close to them (mother, father, significant other, co-workers, siblings). Upon further reflection, we realized that these questions could have been construed by the participants as querying an "other" (e.g., a participant could have interpreted the question "What is your father's first name?" not as a question about themselves but as a question about another person, namely their father). However, this was ambiguous, and thus we could not reclassify these five questions as "Other" questions because some participants may have actually viewed them as "Self" questions. Because this ambiguity was not present for the "Other" questions, we removed the responses to these five questions for each participant, but not for the five corresponding "Other" questions.

Procedure

The procedure was the same as in Experiment 1, except as noted below. After the participant completed informed consent, we administered the MRAB, as in Experiment 1. As before, the investigator remained in the room, behind and out of sight of the participant, for the entire study and answered any questions the participants may have had.

Once the participants completed the MRAB, we presented the deception task with the PsyScope software on a Macintosh computer, beginning with instructions to the participants that they would now answer questions about either themselves or another person with either false or truthful responses. For each trial, the computer screen displayed a cue that instructed the participant how to answer the question that would follow the cue. The cue was either the letter "T," indicating that the participant was to answer the following question with a truthful response, or the letter "F," indicating that the participant was to answer the question with a false response. The four conditions, Self/Truth, Self/Lie, Other/Truth, and Other/Lie, were intermingled. The participants were presented with eight practice questions, two for each of the four conditions. Following the practice trials, the investigator asked the participants whether they had any questions. Once the investigator was satisfied that the participants understood the procedure, the participants were presented with 48 questions, 12 for each of the four conditions. Questions were presented in a random order, except that no more than three in a row could refer to the same person (self vs. other) and no more than five in a row could have the same type of response cue. Participants were asked to respond to each question with a single word.

After the cue had remained on the screen for 4 s, a question was auditorially presented by the computer. Upon presentation of the question, the IRT of the participant's spoken reply was registered by a microphone attached to the computer. Once the participant had responded, a fixation point appeared on the screen to signal that the next cue would appear shortly. The fixation point remained on the screen for 1 s.

Once all of the questions had been presented, participants were asked to rate President Bush on a 10 point scale, with "1" representing the viewpoint "I strongly dislike Bush" and "10" representing the viewpoint "I strongly like Bush." We asked the participants to rate President Bush in order to ensure that our sample did not overwhelmingly differ from the general population in their opinion of President Bush and that emotionality did not play a strong role in their response times. The mean rating for President Bush was 3.95 points (range 1–10 points).

In order to induce the participants actually to lie (that is, to attempt knowingly to deceive someone else), we used the same cover story as in Experiment 1: We told them that they would be paid according to how convincing their lies were, as assessed by an observer in another room (who was supposedly monitoring their face, via the video camera mounted on the computer). In addition, participants were told that the observer would not know which question they were answering. This instruction was given to ensure that participants would not have to be concerned that the observer's knowledge of President Bush would interfere with the success of their deceptive responses.

As before, participants were told that they would win or lose money depending on how convincingly they lied; however, in this study participants were told that every time the observer detected a lie, the participant would lose 25 cents from the hourly wage, and every time the participant succeeded in deceiving the observer, he or she would gain 25 cents. In fact, all participants were paid the full amount as if they had successfully deceived the observer.

Following the computer-administered task, we asked the participants to give us truthful responses to all of the questions they had just been asked. We did this in order to determine whether the participants had correctly followed the cues, which allowed us to compute error rates for each participant in each condition. In addition, we asked the participants to indicate questions for which they did not know the correct response. If a participant did not respond correctly to the cue for a question, or did not know the truthful response, we removed his or her IRT for that question from the sample.

Finally, the investigator then administered the short-form APM for 15 min, debriefed the participant (as in Experiment 1) and paid the maximal possible amount.

Results

For each of the four conditions (Self/Truth, Self/Lie, Other/Truth, Other/Lie), we calculated difference scores for each participant by subtracting the mean IRT for valid truthful responses from the mean IRT for valid deceptive responses.

We began with a two-way analysis of variance, comparing the IRT difference scores for Self versus Other condition for men versus women, and found no effect of gender, $F < 1$; we also did not find any interaction between gender and condition (Wilks' $\lambda = .93$, $F = 2.16$, $p = .15$). Thus, we pooled over gender in all subsequent analyses. Participants took more time to respond to questions in the Other condition, mean: 1358 ms, than in the Self condition, mean: 1167 ms, $t(32) = 2.04$, $p < .005$. Moreover, we did not find a relationship between participants' ratings of Bush and either their mean IRT for the Other condition ($r = -.14$, $p > .1$) or their Other IRT difference scores ($r = .09$, $p > .1$). However, we now found that the participants did in fact take longer to respond with a lie than the truth, and found this in both the Self, lie mean: 1244; truth mean: 1099, $t(32) = 3.05$, $p < .005$, and Other, lie mean: 1441; truth mean: 1281, $t(32) = 2.43$, $p = .02$), conditions.

Next, for each of the nine MRAB tasks, we calculated each participant's RT and ER. We hypothesize not only that these cognitive measures would predict performance on deception tasks, but also that different subsets of these measures would predict performance for different types of lies. And as with Experiment 1, we also included each participant's score on the short-form APM as a means of controlling for general intelligence.

We again used forward stepwise regression analyses to examine the independent contributions of the different MRAB scores, if any, for predicting the IRT difference scores. As in Experiment 1, the mean Raven's score for the participants is again significantly higher ($p < .005$) than the norms reported in Bors and Stokes (1998). Therefore we again performed all analyses both with and without using the Raven's score as a predictor variable. For the first of the two stepwise regressions performed on each condition, we forced the Raven's score in as a predictor as a means of controlling for general intelligence; the second pair of regressions omitted the Raven's score altogether as a independent variable, and only the MRAB scores listed above were used. The results of the stepwise regressions are summarized in Tables 4 and 5.

First, we note that the results change somewhat depending on whether the Raven's score is forced

TABLE 4

Stepwise regression analysis for Experiment 2, with IRT difference scores as criteria and MRAB scores as predictors: Raven's scores forced in at the outset

Criteria	*Predictors*	*β*	*Partial R^2*	*Model R^2*	*F*
Self	Raven's score	−0.37	.17	.17	6.42**
	Spatial working memory RT	0.48	.23	.40	11.45****
Other	Raven's score	−0.23	.00	.00	.02
	Spatial working memory RT	0.37	.24	.24	9.32****
	Mental rotation ER	−0.50	.14	.38	6.79**
	Vigilance RT	0.47	.10	.48	5.59**
	Spatial working memory ER	−0.38	.11	.59	7.23**

Notes: *$p < .10$; **$p < .05$; ***$p < .01$; ****$p < .005$.

TABLE 5
Stepwise regression analysis for Experiment 2, with IRT difference scores as criteria and MRAB scores as predictors: Raven's scores not included

Criteria	*Predictors*	*β*	*Partial R^2*	*Model R^2*	*F*
Self	Spatial working memory RT	0.47	.26	.26	11.06****
	Vigilance RT	0.51	.08	.34	3.67*
	Perceptual reaction time RT	−0.38	.09	.43	4.67**
Other	Spatial working memory RT	0.36	.24	.24	9.64****
	Mental rotation ER	−0.41	.11	.35	5.09**
	Vigilance RT	0.52	.13	.48	7.18**
	Spatial working memory ER	−0.32	.08	.56	5.34**

Notes: *p <.10; **p <.05; ***p <.01; ****p <.005.

into or excluded from the analysis. However, in either case our fundamental point is made: again, we see that different MRAB results predict the different IRT difference scores.

In order to document further that different sets of MRAB results predict different types of lies, we asked whether either of the sets of predictors identified by the stepwise regression analysis could predict both types of lies successfully. Each IRT difference score was fit to both sets of MRAB results with the Raven's score excluded from the analyses. To account for the fact that the two types of lies have differing numbers of MRAB predictors, for each fit we calculated the AIC. The results are presented in Table 6. The set of predictors identified by the stepwise regression in the Self condition (AIC = 454) explains the IRT difference scores for lies about the self better than the set of predictors identified in the Other condition (AIC = 460). Furthermore, the set of predictors identified in the Other condition (AIC = 471) explains the IRT difference scores for lies about President Bush better than the set of predictors identified in the Self condition (AIC = 480). As in Experiment 1, we find that no single set of MRAB scores better predicts both types of lies.

TABLE 6
AIC for IRT difference scores fitted against identified sets of MRAB results for Experiment 2

	IRT difference scores	
MRAB scores	*Self*	*Other*
Spatial working memory RT, Vigilance RT, Perceptual reaction time RT	454	480
Spatial working memory RT, Mental rotation ER, Vigilance RT, Spatial working memory ER	460	471

Discussion

The stepwise multiple regressions where the Raven's scores were forced into the analysis at the outset (thereby removing the contribution of general intelligence) accounted for a larger proportion of overall variance, and thus we will focus on these analyses. We found that three MRAB scores predicted the Other, but not Self, IRT difference scores: ER for the Mental Rotation task, RT for the Vigilance attention task, and ER for the Spatial Working Memory task. For these regression analyses, no MRAB score uniquely predicted the Self IRT difference scores; Spatial Working Memory RT was the only MRAB score that predicted Self lies, but this measure also predicted Other lies. (In contrast, when the Raven's scores were excluded from the regressions, ER on the Perceptual/Motor Control task was a significant predictor for the Self, but not Other, IRT difference scores.) But again, we stress that the important result is that different sets of variables predicted performance for the two types of lies. Moreover, the sets of predictors were not only distinct for the two conditions in this study, but also were distinct from those in Experiment 1.

One puzzle here, however, is that the Filtering task did not predict performance, even though all lies were spontaneous. One possible reason is that knowledge about the self, and even President Bush, is much better learned than knowledge about the work or vacation incident. If so, then participants may not have had as much conflict between the reality and lie, because the two were so easy to distinguish. Furthermore, the spontaneous lies told in this experiment may not have been directly comparable to the spontaneous lies told in Experiment 1. The participants in Experiment 1 were told that their spontaneous lies had to fit into a coherent story. No such instruction was given for Experiment 2. When lying in Experiment 1, participants may have had to inhibit false responses that were not consistent with each other, and such inhibition would not have been necessary in Experiment 2.

It would be useful to conduct the present study using fMRI. If the present interpretation is correct, we would not expect to find anterior cingulate activation in this task.

GENERAL DISCUSSION

An individual's specific basic cognitive strengths and weaknesses should predict his or her performance on various complex tasks that draw on these abilities. For example, performance on a simple mental rotation task should predict performance on a more complicated procedure that requires the use of mental imagery transformations. Using the MRAB, we collected information about individual differences in a wide range of cognitive abilities. We found that these measures did indeed predict performance in four types of deception tasks: producing a memorized set of lies about a life episode; spontaneously creating a coherent set of lies about a life episode; telling a set of isolated lies involving self-knowledge; and telling a set of isolated lies involving knowledge about another person. The fact that variations in performance of different MRAB tasks predicted the time required to produce different types of deception provides evidence that different types of lies arise from distinct cognitive processes.

The present study provided convergent evidence for the conclusions reached by Ganis et al. (2003) in their fMRI study of isolated spontaneous versus memorized coherent lies. In addition, we extended the logic of that study to examine lies about the self versus others—and again found evidence that distinct processes underlie the two types of lies. We also note that lies may differ on more dimensions than the ones considered here. Differences in the subject of the lie may cause different areas of the brain to be accessed; for example, telling a lie about the distant past may require different cognitive processes than telling a lie about more recent events. In addition, lies can differ not only in subject but also in degree. Telling a lie that is a slight exaggeration of the truth may be a cognitively different task than creating a lie that is a complete fabrication.

The lies told in the studies reported here involved unemotional topics; participants were asked to lie in response to questions such as "What is your name?" and "What is your home state?". It is plausible that lying about an emotional topic requires more cognitive effort than lying about a neutral topic. When attempting to deceive another regarding an emotional topic, often the deceiver has to conceal not only the truth, but also the emotions associated with the truth. For example, a man who is trying to deceive his spouse into believing that he is not having an extramarital affair will have to conceal not only the truth about his whereabouts but also any guilt he may have over the affair.

Furthermore, the context surrounding deception may be relevant for the cognitive processes that are evoked. Lies may be told to avoid serious consequences as opposed to being told in a situation where there are only minor consequences to being discovered. Finally, the deceiver's attitude towards the lie may be important. In some situations the deceiver may view telling a lie as being more moral than telling the truth, for example when a lie is told to avoid hurting someone's feelings. This situation may be cognitively different from one in which the deceiver feels that the lie is morally wrong. And so on. These other types of deception may also depend on at least some distinct underlying processes.

Current methods of detecting deception rely on the implicit assumption that deception is a homogenous process with consistent peripheral markers. These methods aim to identify one or more of these peripheral "hallmarks" of deception. One way to circumvent difficulties in recording from the periphery has been to turn to the brain. However, the results of these studies suggest that lying is not a single activity. The fact that successful production of different types of lies depends on different cognitive processes suggests that different forms of deception may be marked not only by different peripheral signals, but also by different patterns of neural activity (Ganis et al., 2003). If so, then an effective, reliable method of deception detection will need to take account of the many dimensions along which lies can vary and the differing cognitive processes that create those lies.

REFERENCES

Akaike, H. (1974). A new look at the statistical model identification. *IEEE Transactions on Automatic Control*, *19*, 716–723.

Berti, A., & Frassinetti, F. (2000). When far becomes near: Remapping of space by tool use. *Journal of Cognitive Neuroscience*, *12*, 415–420.

Bors, D. A., & Stokes, T. L. (1998). Raven's advanced progressive matrices: Norms for first-year university students and the development of a short form. *Educational and Psychological Measurement*, *58*, 382–398.

Burgoon J. K., & Buller, D. B. (1994). Interpersonal deception: III. Effects of deceit on perceived communication and nonverbal behavior dynamics. *Journal of Nonverbal Behavior*, *18*, 155–184.

Bush, G., Whalen, P. J., Rosen, B. R., Jenike, M. A., McInerney, S. C., & Rauch, S. L. (1998). The counting Stroop: An interference task specialized for functional neuroimaging—validation study with functional MRI. *Human Brain Mapping*, *6*, 270–282.

Cody, M. J., Marston, P. J., & Foster, M. (1984). Deception: Paralinguistic and verbal leakage. In R. N. Bostrom (Ed.), *Communication Yearbook 8* (pp. 464–490). Beverly Hills, CA: Sage.

Cohen, J., MacWhinney, B., Flatt, M. & Provost, J. (1993). PsyScope: An interactive graphical system for designing and controlling experiments in the psychology laboratory using Macintosh computers. *Behavior Research Methods, Instruments, and Computers*, *25*, 257–271.

Craik, F. I. M., Moroz, T. M., Moscovitch, M., Stuss, D. T., Winocur, G., Tulving, E., et al. (1999). In search of the self: A positron emission tomography study. *Psychological Science*, *10*, 26–34.

deTurck, M. A. & Miller, G. R. (1985). Deception and arousal: Isolating the behavioral correlates of deception. *Human Communication Research*, *12*, 181–201.

Ekman, P. (1992). *Telling lies*. New York: WW Norton.

Ekman, P. (2001). *Telling lies: clues to deceit in the marketplace, politics, and marriage*. New York: WW Norton.

Farrow, T. F. D., Reilly, R., Towhida, A. R., Herford, A. E., Woodruff, P. W. R., & Spence, S. A. (2003). Sex and personality traits influence the difference between time taken to tell the truth or lie. *Perceptual and Motor Skills*, *97*, 451–460.

Ganis, G., Kosslyn, S. M., Stose, S., Thompson, W. L., & Yurgelun-Todd, D. A. (2003). Neural correlates of different types of deception: An fMRI investigation. *Cerebral Cortex*, *13*, 830–836.

Greene, J. O., O'Hair, H. D., Cody, M. J., & Yen, C. (1985). Planning and control of behavior during deception. *Human Communication Research*, *11*, 335–364.

Gross, C. G., & Graziano, M. S. A. (1995). Multiple representations of space in the brain. *The Neuroscientist*, *1*, 43–50.

Gusnard, D. A., Akbudak, E., Shulman, G. L., & Raichle, M. E. (2001). Medial prefrontal cortex and self-referential mental activity: Relation to a default mode of brain function. *Proceedings of the National Academy of Sciences of the United States of America*, *98*, 4259–4264.

Horvath, F., Jayne, B., & Buckley, J. (1994). Differentiation of truthful and deceptive criminal suspects in behavior analysis interviews. *Journal of Forensic Sciences*, *39*, 793–807.

Iacono, W. G. (2000). The detection of deception. In J. T. Cacioppo, L. G. Tassinary, & G. G. Berntson (Eds.), *Handboook of psychophysiology* (2nd ed., pp. 772–793). Cambridge, UK: Cambridge University Press.

Iacono, W. G., & Lykken, D. T. (1997). The validity of the lie detector: Two surveys of scientific opinion. *Journal of Applied Psychology*, *82*, 426–433.

Johnson, S. C., Baxter, L. C., Wilder, L. S., Pipe, J. G., Heiserman, J. E., Prigatano, G. P. (2002). Neural correlates of self-reflection. *Brain*, *125*, 1808–1814.

Kelley, W. M., Macrae, C. N., Wyland, C. L., Caglar, S., Inati, S. & Heatherton, T. F. (2002). Finding the self? An event-related fMRI study. *Journal of Cognitive Neuroscience*, *14*, 785–794.

Kozel, F. A., Revell, L. J., Lorberbaum, J. P., Shastri, A., Elhai, J. D., Horner, M. D., et al. (2004). A pilot study of functional magnetic resonance imaging brain correlates of deception in healthy young men. *Journal of Neuropsychiatry and Clinical Neuroscience*, *16*, 295–305.

Langleben, D. D., Schroeder, L., Maldjian, J. A., Gur, R. C., McDonald, S., Ragland, J. D., et al. (2002). Brain activity during simulated deception: An event-related functional magnetic resonance study. *NeuroImage*, *15*, 727–732.

McCornack, S. A. (1997). The generation of deceptive messages: Laying the groundwork for a viable theory of interpersonal deception. In J. O. Greene (Ed.), *Message production: Advances in communication theory* (pp. 91–126). Mahwah, NJ: Lawrence Erlbaum Associates.

Mehrabian, A. (1971). Nonverbal betrayal of feeling. *Journal of Experimental Research in Personality*, *5*, 64–73.

National Research Council. (2003). *The Polygraph and Lie Detection*. Committee to Review the Scientific Evidence on the Polygraph. Division of Behavioral and Social Sciences and Education. Washington, DC: The National Academies Press.

Seymour, T. L., Seifert, C. M., Shafto, M. G., & Mosmann, A. L. (2000). Using response time measures to assess "guilty knowledge". *Journal of Applied Psychology*, *85*, 30–37.

Shephard, J. M., & Kosslyn, S. M. (2005). The MiniCog Rapid Assessment Battery: Developing a "blood pressure cuff for the mind". *Aviation, Space and Environmental Medicine*, *76*(Suppl. 6), B192–B197.

Sporer, S. L., and Schwandt, B. (2006). Paraverbal indicators of deception: A meta-analytic synthesis. *Applied Cognitive Psychology*, *20*, 421–446.

Stiff, J., Corman, S., Krizek, B., & Snider, E. (1994). Individual differences and changes in nonverbal behavior: Unmasking the changing faces of deception. *Communication Research*, *21*, 555–581.

Vrij, A. (1994). The impact of information and setting on detection of deception by police detectives. *Journal of Nonverbal Behavior*, *18*, 117–136.

Vrij, A. (2008). *Detecting lies and deceit: Pitfalls and opportunities* (2nd ed.). Chichester, UK: Wiley.
Zuckerman, M., DeFrank, R. S., Hall, J. A., Larrance, D. T., & Rosenthal, R. (1979). Facial and vocal cues of deception and honesty. *Journal of Experimental Social Psychology, 18*, 378–396.

APPENDIX

Here we outline the tasks used in the MiniCog Rapid Assessment Battery. Each of the nine tasks in the battery relies on response times and error rates to assess a particular type of cognitive ability. Together these tasks provide a profile of the participant's attention, working memory, reasoning and perceptual and motor control skills.

Attention

The MRAB contains three tasks that are designed to assess differing types of attention.

Vigilance

The Vigilance task is aimed at assessing an individual's ability to maintain concentration while waiting for a specific event to occur. During this task, the user is given a target stimulus and instructed to press a button whenever that stimulus appears amid other stimuli.

Filtering

The Filtering task is a version of the classic Stroop task. The user is presented with a digit repeated on the Palm Pilot screen multiple times. Users are asked to respond according to how many times the digit appears on the screen (not to the meaning of the digit itself). To perform well on this task, the user must ignore, or filter out, the meaning of the digit. The task contains both "congruent" (i.e., the number of times a digit appears is the same as the meaning of the digit) and "incongruent" (i.e., the number of repetitions is not the same as the meaning of the repeated digit) trials.

Divided attention

The Divided Attention task assesses an individual's ability to focus on two unrelated features of stimuli at once. In this task, the user must attend to both the colors and shapes of stimuli. For example, the user might be asked to press one button if presented with a stimulus that is either black or a triangle (or both) and another button if the stimulus is either white or a circle (or both).

Working memory

The MRAB is designed to assess two types of working memory: verbal working memory and spatial working memory.

Verbal working memory

The Verbal Working Memory task assesses an individual's ability to remember and manipulate verbal information. In this task, the user is presented with a sequence of digits, one at a time. For each digit, the user must determine whether the current digit is the same as one presented two digits earlier in the sequence.

Spatial working memory

The Spatial Working Memory task is designed to assess an individual's ability to remember and manipulate spatial information. In this task, the user is presented with a sequence of digits appearing one at a time in various locations on the screen (the digit itself is otherwise unchanging). For each location, the user must determine whether the current location is the same as the one presented two stimuli earlier.

Reasoning

The MRAB contains three tasks designed to assess reasoning abilities.

Cognitive set switching

The Cognitive Set Switching task assesses an individual's ability to switch between different cognitive sets. On each trial of this task, the participant is presented with four letters. One of the letters differs from the other three in either its case or identity, and the user must identify which letter does not belong. The same criterion is used for a sequence of trials and then changes after a random number of trials, at which point the user must adapt to using the new criterion to determine which letter is out of place.

Three term reasoning

The Three Term Reasoning task assesses an individual's verbal reasoning skills. The user is presented with the first two terms (e.g., "Mary is not shorter than Katherine; Mary is not as tall as Anne") and asked to answer true or false to the third term, which is a question about the implications of the first two terms (e.g., "True or False: Anne is the tallest").

Mental rotation

The Mental Rotation task assesses an individual's spatial reasoning skills. In this task, the user is given a pair of stimuli, with one stimulus rotated relative to the other. The user is asked to determine whether the stimuli are identical or mirror images of each other. To perform this task, the user must be able to mentally rotate the images in order to compare them.

Perceptual and motor control

Finally, the MRAB contains a task to assess an individual's reaction time.

Perceptual/motor reaction time

In the trials of the Perceptual/Motor Reaction Time task, a stimulus appears randomly above one of the four Palm Pilot buttons. The user is asked to press that button as quickly as possible.

SOCIAL NEUROSCIENCE, 2009, 4 (6), 570–574

Psychology Press
Taylor & Francis Group

The effect of deception on motor cortex excitability

Karen J. Kelly
Columbia University, New York, NY, USA

Elizabeth Murray
City University of New York, NY, USA

Veronica Barrios
Montclair State University, Montclair, NJ, USA

Jamie Gorman
Rutgers University-Newark, Newark, NJ, USA

Giorgio Ganis
Harvard University, Cambridge, MA, USA

Julian Paul Keenan
Montclair State University, Montclair, NJ, USA

Although a number of recent neuroimaging studies have examined the relationship between the brain and deception, the neurological correlates of deception are still not well understood. The present study sought to assess differences in cortical excitability during the act of deception by measuring motor evoked potentials (MEPs) during transcranial magnetic stimulation (TMS). Sports fanatics and low-affiliation sports fans were presented with preferred and rival team images and were asked to deceptively or honestly identify their favored team. Hemispheric differences were found including greater excitability of the left motor cortex during the generation of deceptive responses. In contrast to current physiological measures of deception, level of arousal was not found to differentiate truthful and deceptive responses. The results are presented in terms of a complex cognitive pattern contributing to the generation of deceptive responses.

INTRODUCTION

There is a growing literature focused on the functional brain networks involved in deception including malingering (Lee et al., 2005), guilty knowledge (Bunce, Devaraj, Izzetoglu, Onaral, & Pourrezaei, 2005; Langleben et al., 2002; Phan et al., 2005), spontaneous vs. rehearsed lies (Ganis, Kosslyn, Stose, Thompson, & Yurgelun-Todd, 2003), intentional deception (Abe, Suzuki, Mori, Itoh, & Fujii, 2007; Langleben et al., 2005; Mohamed et al., 2006), deception of real vs. imagined events (Abe et al., 2006), and simple vs. complex deception (Lo, Fook-Chong, & Tan,

Correspondence should be addressed to: Julian Paul Keenan, Cognitive Neuroimaging Laboratory, Department of Psychology, 219 Dickson Hall, Montclair State University, Montclair, NJ 07043, USA. E-mail: keenanj@mail.montclair.edu

www.psypress.com/socialneuroscience DOI:10.1080/17470910802424445

2003; Spence et al., 2001). Despite this, a lucid understanding of the cortical processes involved with dishonest responses is lacking.

Recent research has sought to understand the possible mechanisms and processes behind the natural and everyday behavioral occurrence of deception. At the same time, these studies may ultimately contribute to valid and reliable methods for detecting deception for forensic purposes. Deceptive responses appear to be correlated with increased activation of number of different areas including superior medial and inferolateral prefrontal cortex, anterior cingulate cortex (ACC), left motor cortex, occipital cortex, cingulate gyrus, right orbitofrontal, inferior frontal, middle frontal and left middle frontal cortices (Ganis et al., 2003; Kozel et al., 2004b, 2005; Kozel, Padgett, & George, 2004a; Langleben et al., 2005; Marks, Adineh, & Gupta, 2006; Mohamed et al., 2006); indeed, these regions appear to be highly involved across a number of studies (Spence et al., 2004). By focusing on these critical regions, researchers are beginning to develop promising methods for correctly identifying deceptive responses to experimental tasks by specifically observing increases in associated cortical activation, leaving open the possibility of other methods of tracking deception through neurophysiological responses.

Although neuroimaging may soon develop into a useful tool for determining deception, other neurophysiological techniques can contribute to our understanding of cortical activity during the act of lying. While transcranial magnetic stimulation (TMS) has been employed in a 'virtual lesion' manner to disrupt self-deception (Kwan et al., 2007), a more germane use of TMS involves measuring cortical excitability during the production of deception. A study employing this method demonstrated bilateral increased cortical excitability associated with deceptive statements (Lo et al., 2003). This finding was surprising as it was in direct contrast with the prediction that cortical excitability should *not* be greater during deceptive tasks, as verbal and nonverbal indicators of deception reflect inhibition of behavior (Vrij, Edward, Roberts, & Bull, 2000). Furthermore, these researchers found no hemispheric differences, which is at odds with the findings of imaging studies that suggest a reliable left-motor activation during deception (Ganis et al., 2003; Kozel et al., 2005; Langleben et al., 2002; Spence et al., 2001, 2004). Finally, the lies employed, though defined as simple and complex, did not involve material that would personally motivate subjects (i.e., ecological validity).

We therefore sought to examine the precise role of the motor cortex in terms of cortico-spinal excitability as measured by MEPs. To simulate real life deception, we employed sports fanaticism as a model. Participants were required to feign interest in their rival sports team during deception. During these tasks, TMS was delivered and MEPs were recorded to determine whether the left motor cortex was explicitly involved in deception.

MATERIALS AND METHODS

Ten undergraduate males (ages 18–24) participated in exchange for $25 compensation. Six highly identified baseball fans (two Yankee and four Red Sox) and four low-affiliation baseball fans (two Yankee and two Red Sox) provided informed consent and were treated in accordance with the ethical guidelines governing the treatment of human subjects.

All data were collected by and images were presented on Dell computers with Pentium III processors running Acqknowledge 3.7.1 and Superlab 2.0, respectively. A 17-inch Dell monitor was used for the stimuli presentation. Additionally, a chinrest was used to ensure equivalent viewing distance across participants. All photographs of team-relevant images were downloaded from the Internet, and were paired based on approximate content, and size standardized to conform to 2 inch × 2 inch measurement.

Transcranial magnetic stimulation was administered using a TMS-Magstim 200 MonoPulse machine with a figure 8 coil set to motor threshold for each individual. A Biopac Systems MP150 analog amplifier, filtering at a pre-acquisition rate of 1 Hz and 500 Hz, was used to acquire data at 100 ms intervals at 500 Hz. Three electrodes were used to measure motor evoked potentials (one abductor muscle of the thumb, one on the outside line of the thumb, and one ground placed on the back of the hand and approaching the wrist) (Keenan et al., 2001).

The handedness survey and Sports Spectator Identification Scale (SSIS) (Wann & Branscombe, 1993) were administered to each participant prior to enrollment in the study. The SSIS was modified to reflect the specified team allegiance, either Yankee or Red Sox. In addition, the SSIS was modified to reflect Major League Soccer (MLS) affiliation to ensure that control conditions were satisfied in that participants were not strongly identified soccer fans. Cumulative scores on the SSIS were used to distinguish between high- and low-affiliation fans, with those scoring above and below the mean, respectively.

Following the successful location of the motor cortex and motor threshold, participants were provided with written and oral instructions that were formulated according to their team affiliation. In the truthful condition, participants were instructed to respond honestly. In the lying condition, participants were instructed to pretend to be fans of the opposing team. Six practice trials were presented (three lying and three truthful). Following the participant's deceptive response, instructions appeared advising them that their response was being calculated and analyzed. After 30 s, a graph appeared on the screen and participants were informed that the results indicated that they needed to try harder to fool the machine by truly pretending to be a fan of the opposing team while responding.

Each block began with the instruction to tell the truth or to lie. Each trial began with a fixation point at the center of the screen. Following fixation, one image appeared at the top of the screen and its matched pair appeared at the bottom of the screen. Images included team-specific stimuli including: jerseys, players, stadiums, memorabilia, etc. Image location (top or bottom of the screen) and pairs were randomized. Each pair included the favorite team paired with the rival team, the favorite team paired with a neutral team, and the rival team paired with a neutral team. Participants were asked to either truthfully or deceptively identify their preferred team image by indicating the top or bottom position of the image.

DATA ANALYSIS

Scores on the Sport Spectator Identification Scale (SSIS) served as the primary indicator of fan affiliation level. Six participants scored in the high affiliation range and four participants scored in the low affiliation range. MEP data were filtered using a Finite Impulse Response (FIR) linear bandpass filter at 60 Hz and 125 Hz and rectified using an absolute value conversion. Each 100 ms trial was trimmed to exclude the first 24 ms and last 40 ms (resulting in a 35 ms trial) as it was determined that this reflected the most complete MEP across all subjects and conditions.

Following data transformation as described above, a series of ANOVAs were performed to determine differences reflected in the average area under the curve (AUC), peak amplitude (PA), and variability of the MEP data. Three variables were considered: Hemisphere (Right/Left) and Condition (Honest/Deceptive) were included as within-subjects factors. Affiliation (High/Low) was included as the between-subjects factor.

RESULTS

The overall mixed ANOVA (Hemisphere × Condition × Affiliation) was first performed for AUC. It was found that there was no significant three-way interaction, $F(1, 8) = 2.78$, $p = .134$. A significant Hemisphere × Condition two-way interaction was found, $F(1, 8) = 5.15$, $p = .053$. To discover the nature of the interaction, we converted the AUC raw scores to percentage change scores within each hemisphere. It was found that within the right hemisphere there was a 16.10% reduction ($SE = 16.71$) during deceptive responses, while in the left hemisphere there was a 15.70% increase ($SE = 8.74$) when deceptive responses were compared to truthful responses (Figure 1). These data indicated a greater

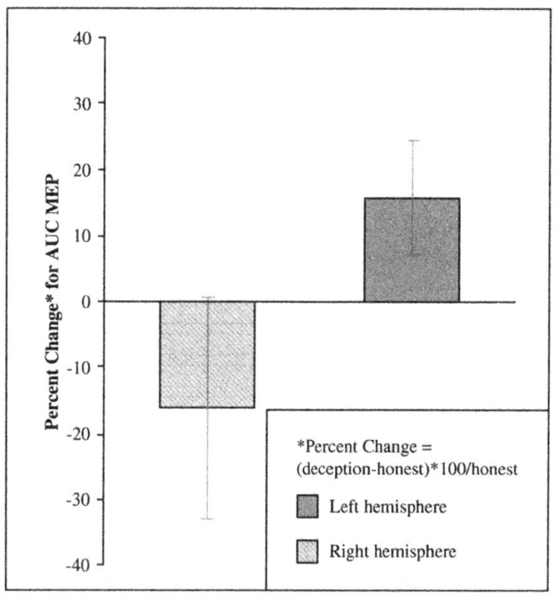

Figure 1. The increased excitability in the left MC during deceptive responses when compared to honest responses, as measured by AUC. In contrast, the right MC exhibited the opposite trend.

excitability for left MC for deceptive responses when compared to honest responses and an opposite trend for the right MC. All other AUC interactions and main effects were nonsignificant with the exception of Affiliation, $F(1, 8) = 6.36$, $p = .036$, in which low-affiliated participants have greater cortical excitability than high-affiliated participants.

A similar $2 \times 2 \times 2$ ANOVA was performed for PA, with no significant interactions or main effects (all p values $> .05$) with the exception of Affiliation, $F(1, 8) = 8.65$, $p = .019$, again with low-affiliated participants having higher average PA than high-affiliated participants. In the final analysis, we examined the variability of response using a similar $2 \times 2 \times 2$ ANOVA. It was found that there was no significant three-way interaction ($p = .373$), but there was a significant two-way Hemisphere × Condition interaction, $F(1, 8) = 5.71$, $p = .044$. The data were recalculated into percentage change scores for each hemisphere (similar to AUC) to simplify the interaction. It was found that the right MC MEPs were 12.58% ($SE = 12.43$) less variable for deceptive statements while the left MC MEPs were 15.24% ($SE = 8.48$) more variable for deceptive statements as compared to honest statements (Figure 2).

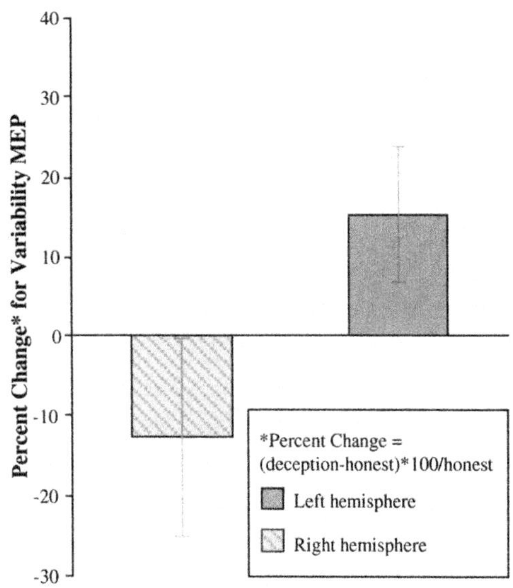

Figure 2. The increased variability in the left MC MEPs during deceptive statements as compared to honest statements. In contrast, the right MC MEPs were less variable during deceptive statements.

DISCUSSION

Although current techniques designed to determine whether an individual is lying rely on physiological responses that reflect arousal, we determined that arousal alone should not be the sole measure of deception. There was no overall (hemispheres collapsed) difference in MEP value when people were lying or being truthful, as reflected in the absence of a main effect of condition, indicating that neither response required more excitability. Further, highly affiliated sports fans evidenced significantly reduced excitability during both the truthful and the deceptive tasks. In spite of the emotional salience of the sports-related images, the self-reported dislike for the rival team, and the overall level of affiliation reported by the sports fanatics which should have resulted in increased arousal (Wann, Royalty, & Rochelle, 2002), participants in the current study were not more aroused and, in fact, demonstrated less arousal during the task (as gauged by both AUC and PA).

There were differences, however, that suggest the need for a broader understanding of the underlying processes contributing to the generation of deceptive responses. The left MC demonstrated significantly increased excitability during deceptive responses, suggesting a possible link between cortical excitability and necessary activation. This is in keeping with fMRI results implicating the left motor cortex during such tasks (Ganis et al., 2003). It can be concluded, therefore, that there is an excitatory component of the deception process that relies on the left MC. Specifically, such differences are seen during deceptive tasks that rely on the use of stimuli and responses that are relevant to the individual's beliefs and preferences.

These findings necessitate a discussion of the role of inhibition. If one does not become more aroused during deceptive statements, perhaps deception is a function of the inhibition of truthful responses (Kozel et al., 2004a,b; Spence et al., 2001). Deception as a primarily inhibitory process explains the lack of excitability during the deceptive task. Further, it is suggestive of hemispheric differences such that one may postulate the role of an inhibitory response generated by the right hemisphere.

Overall, it seems that deception is mediated by a complex pattern of arousal, inhibition, and other factors contributing to the cognitive demands of the task. Successfully measuring and deducing the individual contributions of each of these will aid in the eventual development of more accurate lie detection methods and devices.

REFERENCES

Abe, N., Suzuki, M., Mori, E., Itoh, M., & Fujii, T. (2007). Deceiving others: Distinct neural responses of the prefrontal cortex and amygdala in simple fabrication and deception with social interactions. *Journal of Cognitive Neuroscience, 19*(2), 287–295.

Abe, N., Suzuki, M., Tsukiura, T., Mori, E., Yamaguchi, K., Itoh, M., et al. (2006). Dissociable roles of prefrontal and anterior cingulate cortices in deception. *Cerebral Cortex, 16*(2), 192–199.

Bunce, S. C., Devaraj, A., Izzetoglu, M., Onaral, B., & Pourrezaei, K. (2005). Detecting deception in the brain: A functional near-infrared spectroscopy study of neural correlates of intentional deception. *Proceedings of SPIE, 5769*, 24–32.

Ganis, G., Kosslyn, S. M., Stose, S., Thompson, W. L., & Yurgelun-Todd, D. A. (2003). Neural correlates of different types of deception: An fMRI investigation. *Cerebral Cortex, 13*(8), 830–836.

Keenan, J. P., Nelson, A., O'Connor, M., & Pascual-Leone, A. (2001). Self-recognition and the right hemisphere. *Nature, 409*(6818), 305.

Kozel, F. A., Johnson, K. A., Mu, Q., Grenesko, E. L., Laken, S. J., & George, M. S. (2005). Detecting deception using functional magnetic resonance imaging. *Biological Psychiatry, 58*(8), 605–613.

Kozel, F. A., Padgett, T. M., & George, M. S. (2004a). A replication study of the neural correlates of deception. *Behavioral Neuroscience, 118*(4), 852–856.

Kozel, F. A., Revell, L. J., Lorberbaum, J. P., Shastri, A., Elhai, J. D., Horner, M. D., et al. (2004b). A pilot study of functional magnetic resonance imaging brain correlates of deception in healthy young men. *Journal of Neuropsychiatry and Clinical Neurosciences, 16*(3), 295–305.

Kwan, V. S. Y., Barrios, V., Ganis, G., Gorman, J., Lange, C., Kumar, M., et al. (2007). Assessing the neural correlates of self-enhancement bias: A transcranial magnetic stimulation study. *Experimental Brain Research, 182*(3), 379–835.

Langleben, D. D., Loughead, J. W., Bilker, W. B., Ruparel, K., Childress, A. R., Busch, S. I., et al. (2005). Telling truth from lie in individual subjects with fast event-related fMRI. *Human Brain Mapping, 26*(4), 262–272.

Langleben, D. D., Schroeder, L., Maldjian, J. A., Gur, R. C., McDonald, S., Ragland, J. D., et al. (2002). Brain activity during simulated deception: An event-related functional magnetic resonance study. *NeuroImage, 15*(3), 727–732.

Lee, T. M., Liu, H. L., Chan, C. C., Ng, Y. B., Fox, P. T., & Gao, J. H. (2005). Neural correlates of feigned memory impairment. *NeuroImage, 28*(2), 305–313.

Lo, Y. L., Fook-Chong, S., & Tan, E. K. (2003). Increased cortical excitability in human deception. *NeuroReport, 14*(7), 1021–1024.

Marks, D. H., Adineh, M., & Gupta, S. (2006). Determination of truth from deception using functional MRI and cognitive engrams. *Internet Journal of Radiology, 5*(1).

Mohamed, F. B., Faro, S. H., Gordon, N. J., Platek, S. M., Ahmad, H., & Williams, J. M. (2006). Brain mapping of deception and truth telling about an ecologically valid situation: Functional MR imaging and polygraph investigation—Initial experience. *Radiology, 238*(2), 679–688.

Phan, K. L., Magalhaes, A., Ziemlewicz, T. J., Fitzgerald, D. A., Green, C., & Smith, W. (2005). Neural correlates of telling lies: A functional magnetic resonance imaging study at 4 tesla. *Academic Radiology, 12*(2), 164–172.

Spence, S. A., Farrow, T. F., Herford, A. E., Wilkinson, I. D., Zheng, Y., & Woodruff, P. W. (2001). Behavioural and functional anatomical correlates of deception in humans. *NeuroReport, 12*(13), 2849–2853.

Spence, S. A., Hunter, M. D., Farrow, T. F., Green, R. D., Leung, D. H., Hughes, C. J., et al. (2004). A cognitive neurobiological account of deception: Evidence from functional neuroimaging. *Philosophical Transactions of the Royal Society of London B: Biological Sciences, 359*(1451), 1755–1762.

Vrij, A., Edward, K., Roberts, K. P., & Bull, R. (2000). Detecting deceit via analysis of verbal and nonverbal behavior. *Journal of Nonverbal Behavior, 24*, 239–263.

Wann, D. L., & Branscombe, N. R. (1993). Sports fans: Measuring degree of identification with their team. *Journal of Sport Psychology, 24*(1), 1–17.

Wann, D. L., Royalty, J. L., & Rochelle, A. R. (2002). Using motivation and team identification to predict sport fans' emotional responses to team performance. *Journal of Sport Behavior, 25*(2), 207–216.

Subject index

A' 498–499
Akaike Information Criterion (AIK) 560
Amygdala 521, 522, 523, 524
Anterior cingulate cortex (ACC) 469, 516, 529, 530, 532, 534, 535, 548, 550, 555, 556, 571
Arousal 466, 474, 511, 539
Attention 556
Attitude towards lying 566

Bayesian classification 478–479, 481, 485
Behavioural cues 465
BOLD measures 469, 511, 521, 529, 530, 543
Bootstrapping 479, 496–497, 498, 505, 506
Brain fingerprinting 536

Caudate 515, 529, 532, 534, 535
Cerebellum 545, 546, 549
Cingulate 541, *see also* Anterior cingulate cortex
Cingulate gyrus 545, 546, 547, 549, 571
 medial 532, 535
Cingulate-insular-prefrontal networks 470, 547, 550
Claustrum 545, 546, 547, 548
Cognitive abilities 556, 560–562, 564–566
Commercial gains 536
Concealed Information Test (CIT) 467–468, 492
Context 566
Control-question technique (CQT) 474
Countermeasures 468, 492, 525, 537
Court testimony 468
Cuneus 545, 546, 549

Deception
 definition 465
 process 466
Deese-Roediger-McDermott (DRM) paradigm 468, 475, 486
Differentiation paradigm 468
Dorsolateral prefrontal cortex (DLPFC) 529, 532, 534, 535

Electrodermal activity (EDA) 469, 511, 512, 513, 514–517
Employment screening 519
Ethical issues 519, 536
Event-related potentials (ERPs) 466, 468, 540
 false memory 468, 474, 475, 476, 478–482, 483–487
 scalp distribution 468, 491–492, 494, 495–496, 497–498, 499–508
Exploratory analysis 478, 480–481, 484–485, 486

False memory 468, 474–476, 480–482, 484–487
Frontal cortex
 inferior 522, 571
 medial 467
 middle 571

Frontopolar region 532
Functional magnetic resonance imaging (fMRI) 466, 469–470, 529–537, 540
 electrodermal activity and 469, 511–513, 514–517
 self- and other-related lies 470, 540–541, 542–544, 545–550
 towards clinical trials 518–526
Fusiform gyrus 529, 545, 546, 549
Fuzzy trace theory 487

General linear model (GLM) 521
Gist information 487
Group data 466
Guilty Knowledge Test (GKT) 467–468, 474, 530, 536–537

Heschl's gyrus 515
Hippocampus 521, 522, 523, 524, 545, 546, 547, 548–549

Imagery 549, 561, 566
Individual differences 467, 469, 470, 539–553, 556
Inferior frontal gyrus 521, 522, 523, 524, 530, 532, 534, 545, 546, 548
Inferior frontal lobe 522, 571
Inferior occipital gyrus 545, 546, 549
Inferior parietal lobe 521, 522, 523, 524, 529, 534, 547, 550
Inferior temporal gyrus 515, 522, 524
Inhibition 573
Initial response time, (IRT) 555, 556, 559–562, 564, 565
Instructions to lie 467
Insula 515, 541, 545, 546, 547, 548
 cingulate-insular-prefrontal networks 470, 547, 550
Interference 470, 471, 541, 548, 550

Laboratory versus field studies 467
Legal issues 468, 519, 536
Limbic system 519, 521, 522, 524–525
Lingual gyrus 532, 535, 545, 546, 549

Medial cingulate gyrus 532, 535
Medial frontal cortex 467
Medial frontal gyrus 521, 522, 523, 524, 529, 532, 534, 545, 546, 549
Medial frontal negatives 466
Medial prefrontal cortex (mPFC) 469, 470, 529, 532, 534, 536
Memory 468, 474
 false memory 468, 474–476, 480–482, 484–487
 working memory 519, 556, 560, 561, 564, 565
Mental imagery 549, 561, 566
Mental rotation 556, 560, 561, 564, 565, 566
Middle frontal cortex 571

Middle frontal gyrus 530, 532, 534, 535, 545, 546, 547, 548
Middle occipital gyrus 545, 546, 549
Middle temporal gyrus 515, 521, 522, 523, 524, 532, 534, 535
Mind-reading 525
MiniCog Rapid Assessment Battery (MRAB) 556, 560, 564–566, 568–569
Minimally sufficient processes 541
Mock crime scenario 493
Monte Carlo simulations 521, 531
Moral dilemmas 525
Motivation 467
Motor cortex excitability 470, 571, 573
Motor evoked potentials 571, 572–573
"Multiple issue" question 492, 507

N1/N2, 480, 484

Occipital lobe 545, 571
Oddball probability 492, 507
Olfactory area 515
Operculum 515
Orbitofrontal cortex 469, 515, 516, 571
Other-related lies 470, 540–541, 542–550, 562–566

P1, 480, 484
P2, 480, 484
P300 (P3) 466
  countermeasures 468, 492
  false memory 468, 475, 476, 478, 479–480, 483–484, 486–487
  scalp distribution 468, 491–492, 494, 495–496, 497–498, 499–508
Paracentral lobule 515
Parahippocampal gyrus 521, 522, 523, 545, 546, 547, 549
Parietal lobe 529, 545
  inferior 521, 522, 523, 524, 529, 534, 547, 550
  prefronto-parietal systems 519, 522, 525
  temporo-parietal junction (TPJ) 532, 534, 535
Pathological lying 549
Polygraphs 474, 511, 528–529, 554
Postcentral gyrus 515
Precentral gyrus 515
Precuneus 515, 545, 546, 547, 548, 549
Prefrontal cortex 529, 541, 548, 550, 571
  cingulate-insular-prefrontal networks 470, 547, 550
  dorsolateral (DLPFC) 529, 532, 534, 535
  medial (mPFC) 469, 470, 529, 532, 534, 536
  ventrolateral (VLPFC) 529, 530, 532, 534, 535
Prefronto-parietal systems 519, 522, 525
Prepared lies 467, 470, 555–562, 566
Prepotent responses 466, 470, 541
Priming, semantic 486
Principal component analysis 497
Pseudo-ROIs 532, 533

Rate-limiting processes 470, 541, 548–549, 550
Rehearsed lies 467, 470, 555–562, 566
Relationship management 465
Resolution 471, 541, 548, 550
Resumé Query Test (RQT) 469, 519–522, 525
ROC curves 482, 485, 530

Scaffolding 550
Self-related lies 470, 540–541, 542–550, 562–566
Semantic priming 486
Set-switching 556, 561
Single individual data 466
Slow wave 487
Social inappropriateness 525
Social relationships 465
Source monitoring 487
Spatial working memory 556, 560, 561, 564, 565
Spontaneous lies 467, 470, 555–562, 566
Sports Spectator Identification Scale (SSIS) 571–572
Stroop task (filtering), variant 556, 560, 561, 562, 565
Subcallosal gyrus 522, 523
Subcortex 529
Superior frontal gyrus 534, 545, 546, 547, 548, 549
Superior medial frontal gyrus 532
Superior temporal gyrus 515, 521, 522, 523, 534
Superior temporal sulci 529
Supplementary motor area 515
Supramarginal gyrus 515, 521, 522, 523

t-test maps 478, 481, 484–485, 498
Temporal lobe 529, 545
Temporo-parietal junction (TPJ) 532, 534, 535
Thalamus 545, 546, 549
Theory of mind 525
Transcranial magnetic stimulation (TMS) 470, 571

Uncus 521, 522, 523

Varimax rotation 497–498
Ventrolateral prefrontal cortex (VLPFC) 529, 530, 532, 534, 535
Verbal working memory 556, 560, 561
Vigilance 556, 560, 561–562, 564, 565
Visual cortex 529

Working memory 519, 556, 560, 561, 564, 565

For Product Safety Concerns and Information please contact our EU
representative GPSR@taylorandfrancis.com Taylor & Francis Verlag GmbH,
Kaufingerstraße 24, 80331 München, Germany

Printed and bound by CPI Group (UK) Ltd, Croydon, CR0 4YY
21/10/2025
01981231-0015